LABORATORY EARTH

LABORATORY EARTH

The Planetary Gamble We Can't Afford to Lose

STEPHEN H. SCHNEIDER

A Member of the
Perseus Books Group

The Science Masters Series is a global publishing venture consisting of original science books written by leading scientists and published by a worldwide team of twenty-six publishers assembled by John Brockman. The series was conceived by Anthony Cheetham of Orion Publishers and John Brockman of Brockman Inc., a New York literary agency, and developed in coordination with Basic Books.

••••••••••••••

The Science Masters name and marks are owned by and licensed to the publisher by Brockman Inc.

••••••••••••••

Published by Basic Books,
A Member of the Perseus Books Group.

••••••••••••••

••••••••••••••

Library of Congress Cataloging-in-Publication Data

Schneider, Stephen Henry.
    Laboratory earth : the planetary gamble we can't afford to lose / Stephen H. Schneider. — 1st ed.
        p.   cm.
    Includes index.
    ISBN 0-465-07279-8 (cloth)
    ISBN 0-465-07280-1 (paper)
    1. Global environmental change.   2. Climatic changes.   3. Nature—Influence of human beings on.   I. Title.
GE149.S36   1997
304.2'8—dc20   96-42987
                                                      CIP

••••••••••••••

97 98 99 00 01 ❖/RRD 10 9 8 7 6 5 4 3 2 1

CONTENTS

It's a cliché, I suppose, to assert that by the time you can finally afford to purchase or accomplish some urgent dream of youth, you no longer crave it. Clearly, our perspectives change with time. In my school years, weeks of dread preceded an assignment to write a "long" paper of some twenty pages. Decades later, this ink-stained veteran of perhaps twenty thousand pages faced months of a greater anxiety: trying to cram the essence of the very complex science, technology, and policy controversies surrounding the subject of global environmental change into a "short" book of some hundred and fifty pages.

I am indebted to my agent, John Brockman, not only for reminding me that such an exercise in brevity increases the reach of one's ideas, but for creating the Science Masters series, which allows people an opportunity to be introduced in moderate depth to the substance and implications of the critical scientific issues of the day. Trading off content and context, scientific issues against policy controversies, and third-person reporting versus first-person advocacy—all in an accessible, compact package—was a major challenge. Of necessity, compromises were made; but a number of end-notes and additional readings are offered for each chapter (including some of my own writings, so that any reader who may want to hear me defend in more detail some of the things asserted in these pages has citations to pursue).

Despite a valiant effort, my early drafts were too long and sometimes too diffuse. Editorial comments and scientific critiques from Jerry Lyons, Jacques Grinevald, Stuart Pimm, Russell Burke, Larry Goulder, and Richard Manning were helpful in pointing out such problems. And I am grateful to the science writer Joel Shurkin for agreeing to take up the editorial surgeon's knife. After his (occasionally painful but) skillful operation, the readers have the benefit of a more logically organized, compact, and accessible work. I appreciate the perspectives on the human psyche I gained from Sharon Conarton, since I firmly believe that it will be difficult to solve dimly perceived global problems when too many of us are steeped in too much personal denial. I also wish to thank Debra Sacks for efficient word processing of several drafts and cheerful accommodation to demanding deadlines at several stages, and Katerina Kivel for editing.

My children, Rebecca and Adam, had to deal with a father who often would appear at breakfast bleary-eyed from late hours of writing and editing, only to be reminded by them that I had insisted only hours before that *they* get a good night's sleep to keep healthy and alert; somehow this intrepid author/father had strangely missed his own advice. Their support of my absorption in this project is lovingly accepted. And for Terry Root, my partner in both professional and personal spheres of life, I appreciated her timely and credible perspectives when my judgments were squeezed from too much internal pressure. Even more so, I am grateful for her choice not to add external pressure when I was recharging and perhaps looking to be nudged out of idling. We need the freedom to accomplish our art at the pace we negotiate with ourselves, and that insight she helped to reveal. In any case, this book is the product of that process. It is offered in the hope that at least some readers will be motivated to pursue their knowledge of the Earth much further, and that nearly all will resolve to get involved in being part of the solution to Earth system problems.

INTRODUCTION

· ·

IT'S A MATTER OF SCALE

Remember the famous photographs the astronauts took in space in the late 1960s that transformed global consciousness about the Earth? White clouds swirled around a blue globe with white ice caps and reddish deserts. The spiral patterns of storms stood out as bold features occupying regions the size of the New England states—1,000 kilometers or so in scale. That's one way of looking at the atmosphere. An airplane passenger on a turbulent flight might think the atmospheric action is at the scale of hundreds of meters as the plane is tossed about in the sky. A balloonist who can see individual rain droplets or snowflakes leisurely drift by might conclude that the atmosphere must be understood at the microscale of millimeters. These observations are all "right" in a sense. It depends on what you are looking for, or at.

We might look up at a stormy sky, for example, and see clouds drifting from east to west. Does that mean the storm overhead is moving from east to west? Suppose the satellite map on the weather news that evening on TV shows us that although the local winds circulating at that instant were indeed moving from east to west, the overall storm was actually moving from west to east? There was nothing wrong with our local observations, just with our larger-scale hypothesis. We needed a bigger picture to get the large-scale relationships right. Or, as the mathematical ecologist Simon Levin of Princeton University once put it, the world looks very different, depending on the size of the window you are looking through.[1]

The problem of seeing the world at one scale and extrapolating that observation to make judgments at other scales is at the root of more unnecessary contention than just about any practice I can think of—in interpersonal relationships as much as in arcane scientific debates.[2]

Nature shows amazing richness in its range of spatial scales and their interactions over what phenomena occur. Richness occurs over a range of time scales as well. You know from experience that winds blow and oceans flow, but those aren't the only parts of the Earth that are dynamic. Our "solid" Earth is not solid, not forever fixed on the map in space and time. In fact, the land moves about, in response to natural forces. The drift of continents, we'll see later, can have a major influence on the climate and on life.

Except for local phenomena like earthquakes or landslides or glaciers, whose motions are observable in human time frames, the time scales for major, continental-scale Earth motions are thousands to millions of years. It took special tools and creative insights to see these motions. How the "solid" Earth interacts with air, water, and life is important for understanding the Earth as a system.

Knowledge of cloud microphysics, even in great depth, will not by itself provide much context about the behavior of the Earth's weather machine as viewable from space at large scales. So at what scale should we focus our discussions of weather, climate, ecology, society, and environmental change?

Our own experience isn't enough—our personal scale is too limited—to see the full range of important phenomena in nature. We need the observations and inferences of a larger community—Earth system scientists, in this example—to open our perceptual window to the rich variety of nature that surrounds us.

CONTENT WITH CONTEXT
....

There has long been a fundamental tension among those who argue that without in-depth study our analyses will be shallow. Indeed, specialization has marked both academic and economic success stories since the Industrial Revolution. But, increasingly, there are those who say that without some sense of the broad context of real problems, disciplinary specialization may not provide what is needed to understand or solve pressing issues. To me, it is not sensible to debate for long whether it is worse to approach real problems only from the narrow, but deep, purview or to deemphasize sharply focused depth and instead stress integration across specialized disciplines. (The latter sometimes imposes career risks for interdisciplinarians, since problem solving often means fashioning originality at the intersection of disciplines, with not enough disciplinary originality to constitute a "respectable" contribution in that narrow specialty.) Despite the passion on both sides, content versus context (or large scale versus small scale) is a foolish, false dichotomy. The world obviously needs both large- and small-scale views with enough content to avoid being superficial, blended with sufficient context to address demanding, real-world problems.

Although space limitations prevent me from discussing here all relevant fields in the depth that a specialized account would demand, I will explore a wide range of environmentally relevant topics in sufficient detail to explain much of what is known about climate change and its ecological and societal implications. I will also identify what is speculative about climate change in the overall environmental debate. I'll use the context of practical environmental and economic trade-offs to help guide the selection of a representative set of content areas.

Awareness that pollution can degrade our environment is hardly new. That was dramatically learned centuries ago in the era of uncontrolled coal burning that fueled the infamous London smogs. Centuries earlier, soil erosion on denuded hillsides in Asia taught a painful lesson of the need to farm or deforest with careful conservation practices. But these early lessons had two characteristics: they were discovered at local or regional scales, and they were learned after the fact—once the damage was apparent. The twenty-first century environmental problems are unique because the scale is truly global rather than simply local to regional. Even more serious, potentially long-lasting, even irreversible, effects are quite possible, thus it is no longer acceptable simply to learn by doing. When the laboratory is the Earth, we need to anticipate the outcome of our global-scale experiments *before* we perform them. At least that is the rationale undergirding the Earth systems science we'll explore here.

The planetary-scale environmental issues I'll address in these pages have come to be called "global change." That phrase was invented by people who study the Earth as a whole system, to refer to the changes on a global scale that affect Earth systems (physical, biological, and social) that are interconnected and for which humans have some role in effecting those changes. Why study continental drift as part of "global change," since humans are certainly not able to make continents drift? Because if we don't understand how drifting continents affect the gases in the atmosphere, the climate, or biological evolution, then we won't have the background knowledge necessary to forecast credibly so-called anthropogenic (that is, human-induced) sets of global changes.

I'll touch on knowledge from traditional academic disciplines such as geology, ecology, atmospheric science, biology, energy technology, chemistry, agronomy, oceanography, political science, economics, and even psychology.[3] I'll also be looking at how humans are disturbing various components of the planetary system. In the course of the book a number of Earth systems science questions will be addressed:

- How long did it take for the climate and life to evolve this far?
- How does the Earth work as a coupled set of subsystems that includes living and nonliving parts?
- How are people disturbing the Earth system?
- What have we learned from the workings of the natural system that can help us forecast how human disturbances might affect it?
- What are some of the trade-offs between environmental protection and economic development, and how can both sets of these seemingly conflicting interests be reconciled?

THE WHOLE MAY BE WORSE THAN THE SUM OF ITS PARTS
....

One of the most potentially serious global change problems is the combined or synergistic effects of habitat fragmentation and climate change. People fragment natural habitats for farmland, settlements, mines, or other development activities. If the climate changes, individual species of plants and animals will be forced to adjust if they can, as they have in the past.[4]

Typically, they'll migrate, as spruce trees did when the last ice age waned some ten thousand years ago. But the landscape has changed dramatically since then. Could all the migrating species that survived the ice age make it across the freeways, agricultural zones, industrial parks, military bases, and cities of the twenty-first century? Good science is necessary to help answer how such biological conservation practices can take place in the most economically efficient or politically practical ways. Global change science will involve looking at these kinds of questions. To answer them, we need to go to academic disciplines and ask, What knowledge do you have? The two most important questions to ask specialists, whether medical doctors or Earth system scientists, are simply: What can happen? and What are the odds of it happening?

The Earth systems scientist tries to integrate the information from many disciplines in an original synthesis that addresses real problems at the scales at which they occur.

WE HAVE MET THE ENEMY

People rarely intend to create environmental problems (illegal toxic-waste dumping and igniting the oil fields of an invaded country being some exceptions). Rather, most environmental ills simply emerge inadvertently from the sum of myriad, seemingly minuscule individual actions occurring at small scales, but around the globe. Whether by accident or design, the locally poisoned fish or globally altered climate is, nonetheless, damaged. Motive is irrelevant to environmental impacts. It applies only to dealing with the aftermath. Much of what we do to the environment is an experiment with Planet Earth, whether we intend it to be or not. It is everyone's job to make the unintended potential consequences of our behavior conscious—even if ignorance or denial is the politically simpler "solution." As the Stanford University population biologist Paul Ehrlich once quipped: "Ignorance of the laws of nature is no excuse."

THE HUMAN DIMENSION
....

Causes of global-scale environmental degradations are most often ascribed to increasing numbers of people demanding higher standards of living and using technology or practices that pollute or fragment the landscape. There is an equation for this, formulated in 1971 by Paul Ehrlich and the then University of California at Berkeley energy analyst John Holdren: $I = PAT$.[5] That is, environmental impact (I) equals population (P) times affluence per capita (A) times technology used (T).

When an observer abandons the large or global scale and looks instead at local environmental problems, these three factors may not be seen as easily. Different factors are identifiable at different scales. When viewed at the local level, you may find that corrupt officials or unaccountable industries stand out as prime causes of environmental problems. At larger scales, the problem may appear to be increasing use of land or energy and burgeoning populations.

I said we cannot avoid the human dimensions of global change if our analyses are to be useful. Some nations are economically better off than others, and more economic equity is a driving force in economic planning in the less developed nations. Tensions erupt between nations when it is asserted that those plans could threaten the global environment. At a local level, taxes on polluting fuels are an incentive for conservation and the development or deployment of cleaner alternatives. But taxes raise the price of energy, which has a greater impact on poor people than wealthy ones. People facing economic hardships usually have their priorities focused on economic growth more than on environmental protection. Such environment/development or equity/efficiency trade-off issues are already in the news and will lead to major debates in the decades ahead. So too is the problem known as "intergenerational equity": The desire for economic progress today, and the wish to leave our heirs richer than ourselves, may boomerang and leave a legacy of environmental problems to later generations who cannot participate in today's decision making.

There are between five and one half and six billion people in the world today, with one billion living on the margins of nutritional deprivation and tens of millions who die every year from preventable illnesses related to malnutrition. These people demand and deserve improved standards of living, but decisions made to satisfy that right cannot be just if they ignore the effects on the Earth. Even the grounds on which these debates are being held is disputed. The social scientists Robin Cantor and Steve Rayner noted that, like other human value conflicts, "the environmental debates can be understood in the context of people invoking different mythologies of the workings of nature to support their various political and moral beliefs."[6] Thus, natural and social sciences need to be blended with humanistic studies to fully illuminate the value dimensions of environment/development dilemmas. As we increase our understanding of the systems that control Earth's environment, the myriad interconnections and potential solutions will crystallize.

I will discuss both the local problems and their impact on the global environment and global problems, which can in fact affect the local environment. Studying the environment fascinates because everything is linked to everything else in the system we call Earth, and while the connections among variables can be subtle, the effects sometimes are all too obvious. We obviously don't know all the answers yet—and not even all the important questions! It will take interdisciplinary teams many decades to adequately assess the science and management problems of global change. But a great deal is already known, and much can be done to reduce risks.[7] An informed public with the scientific knowledge and political will to make a difference can deal with many of the difficult questions that confront us. It is to that end that this book is dedicated.

But before we can peer into the shadowy future of climate and life, it is essential to journey back to our biogeographical roots: the Archean era of the distant past, when a young Earth was first beginning to nourish life.

....

CHAPTER I

THE ORGANIC AND NONLIVING EARTH: A DYNAMIC COHESION

I doubt there is an earth scientist alive who wouldn't jump at the chance for a trip on a time machine, to clock and measure the changes that naturally occurred on Earth eons ago. The scientist could move across the millennia, watching the plates on which continents ride slide across the surface of the Earth, altering not only their positions and the atmospheric composition but the biological life they carry as passengers. The scientist could monitor changes in the air, land, and waters that influence the evolution of life and, with proper attention, could detect how life, in turn, changed the character of air, land, and waters. The organic and the inorganic are connected: geochemistry and biology, geology and climatology. In time-machine scale, everything is in motion, constantly changing, like a giant, intricate, and evolving web of living and inanimate parts that together form a dynamic cohesion. But this pattern is not easily grasped by an observer without benefit of this marvel of fiction, unless she or he is part of a community of the curious using refined methods to reveal the immense patterns that emerge across the eons. That community and its methods are, of course, what today we call Earth systems science.

This dynamism plays out over geological time, an almost unimaginable span in which a thousand years is a wink. It usually takes many such winks to attract a geologist's attention. The character in H. G. Wells's *The Time Machine* could see the evolution of civilization over centuries; a biologist or geologist or climatologist in a sturdier device traveling back over a far longer period could watch the evolution of organisms and their interrelationship with the Earth they inhabit.

One period especially interesting to visit would be the era of the dawn of life, some 3.5 billion years ago, during the so-called Archean Age. We might be able to solve a sovereign scientific mystery that not only encapsulates Earth systems science but is at the core of the modern debate over global warming and the dangers of some of our unintended experiments with our world. What would we find there?

We would likely see a sun rising behind clouds in the sky, tall, smoking volcanoes, and sea waves lapping onto a treeless, grassless, barren plain. Strange-looking, meter-wide, mushroomlike rocks stand at the shoreline. We dare not venture outside our time machine without eye and skin protection because of dangerously high levels of ultraviolet radiation, levels high enough to threaten the long-term survival of any known life forms on land or in the air. We also must wear oxygen masks, for the atmosphere is composed primarily of carbon dioxide. There is some oxygen, but at only a billionth or so of the levels we know today.

The air temperature is hot—38°C (100°F)—but the noonday sun somehow appears dimmer and slightly smaller than the one we're used to back in the Holocene interglacial, our time. Solar panels outside our time machine read about 600 watts of incoming energy, about 25 percent less than the power we now receive from the sun. Three and a half billion years ago, the sun *is* smaller than our sun.

But why? When nuclear physics is applied to solar processes, it suggests that our star, like most of its type, grew fatter and brighter as its thermonuclear reactors converted hydrogen to helium. Most scientists believe that the solar luminosity has increased about 30 percent since the formation of the Earth some 4.5 billion years ago, 5 percent of that in the past 600 million years. This was the era of rapid biological evolution that left its indelible fossil footprints in the rocks we dig up today.

A SUPER GREENHOUSE EFFECT

Most climatologists would unhesitatingly say that a cut in solar heat input to the Earth by some 25 percent today would plunge us into a deep freeze. But the Archean was apparently warm, not frozen—remember, the outdoor thermometer of our time machine read a toasty 38°C.

This dilemma is popularly known as the "faint early sun paradox." In 1970, Carl Sagan and George Mullen, at Cornell University, proposed one solution to the paradox: a super greenhouse effect.[1] They suggested that two gases, methane and ammonia, are very effective at trapping infrared radiation in the lower part of the Earth's atmosphere, and that these may have been present in sufficient abundance in the Archean to make up for the solar deficit and keep the climate temperate.

Critics called their idea fanciful, arguing that these gases are highly reactive, have short lifetimes in the atmosphere, and therefore need to be continuously replenished (presumably by life). If so, how could methane and ammonia have been present in large enough concentrations to make the Earth warm enough to get life going? We don't know—one reason why that time machine is such an attractive fantasy for the Earth-curious.

This issue of whether methane (CH_4) and ammonia (NH_3) were produced by biological processes or by processes that had nothing to do with biological forms in the Archean era is still unresolved, but Sagan's and Mullen's basic idea is accepted by most scientists. However, today's variant suggests that carbon dioxide (CO_2) may have been the main super greenhouse gas, rather than CH_4 or NH_3. The shadow of this theory looms over us today. If such a phenomenon occurred in the Archean time, couldn't it happen again?

To answer that crucial question, we have to be cognizant of the processes that affect the composition and structure of the atmosphere.[2]

In science, increased understanding does not always mean increased certainty, at least at first. The solution to one problem frequently raises another problem. In this case: If some hundred times greater than present concentrations of CO_2 kept the Archean balmy, what happened to keep the climate from dramatically overheating during the next 3 billion years as the sun brightened by 25 percent or so?

Answers (hypotheses, really) to this dilemma typically take two (sometimes conflicting) forms: One theory says that temperatures and CO_2 were controlled by inorganic geochemical processes that removed CO_2, the other that they were controlled by biological removal of CO_2. Or both. Either way, each theory is based on a process called *negative feedback*.

All of us endothermic (warm-blooded) creatures contain stabilizing, negative feedback mechanisms. We are what physiologists call homeostatic systems. If we get too hot, we sweat to cool down a negative or stabilizing feedback. If we get too cold, we shiver, which is a mechanical way of cranking up our metabolic level and generating heat, which is also a stabilizing feedback.

There are many feedback processes in the climatic system, some of them stabilizing, like a thermostat, but some of them destabilizing. If the Earth warms up, for example, what happens to the snow and ice? Some melts, and this melting ends up replacing bright, white, reflective stuff with green trees or brown dirt or blue oceans. They're darker than the snow fields and so are going to absorb more sunlight. If we could warm up the Earth somehow and thereby induce some snow to melt, the Earth would then absorb more sunlight and this feedback process would accelerate the warming. That's a positive feedback. But if warming causes more water to evaporate and this makes whiter clouds which reflect more sunlight back to space, reducing the heating of the planet, that's a negative feedback.

Let's return to the debate about CO_2 removal. For the model that geochemical processes controlled CO_2 in the atmosphere, James Walker, Paul Hays, and James Kasting, all then at the University of Michigan, proposed in 1980 a weathering-climate stabilizing feedback system referred to by their colleagues as WHAK (an acronym for their initials).[3] They suggested that as the climate warmed, more water evaporated and the hydrological cycle became more vigorous with increasing precipitation and runoff.

The WHAK mechanism operates on time scales of tens to many hundreds of millions of years. It was not intended by itself to describe the shorter-term variations in CO_2 that may help explain the extreme warmth of the peak of the dinosaur era or the extreme cold of the last ice age, some twenty thousand years ago. (More on that later.)

Given the high concentrations of CO_2 in the atmosphere, CO_2 mixing with rainwater creates carbonic acid runoff; increased precipitation would expose minerals at the surface to larger amounts of this weathering liquid. In the weathering process that WHAK invoked, minerals such as calcium and magnesium silicates would combine with carbon in the atmosphere, drawing down the CO_2 concentration in the air and locking that carbon in sedimentary rocks such as calcium carbonate (limestone) and magnesium carbonate (dolomite). Less CO_2 in the atmosphere means less greenhouse effect, thereby offsetting the increase in solar luminosity over geological time through this inorganic negative feedback process.

IS GAIA REAL?
....

The second class of theories for getting rid of high Archean levels of atmospheric CO_2 as the sun grew more luminous involves removal through a biological process. One idea for a biological negative feedback mechanism was advanced by the British scientist and author James Lovelock.[4] He attempted to explain how life could act at a global scale as an automatic negative feedback control system—what he termed the "Gaia hypothesis," after the Greek goddess of the Earth, at the suggestion of his neighbor, the author William Golding. The Gaia hypothesis, which was not taken very seriously by scientists at first and still has many critics,[5] states that Earth's atmosphere is an integral, regulated, and necessary part of life itself, and that, for thousands of millions of years, life has controlled the temperature, chemical composition, oxidizing ability, and acidity of Earth's atmosphere. Life wields active control over the planet's environment, Gaians assert. Lovelock based one mechanism on the assumption that photosynthesizing microorganisms, like phytoplankton, would be very biologically productive in a high CO_2 environment, and that they could rapidly (but still on geological time frames) remove CO_2 from the air and oceans, converting it into calcium carbonate sediments as they died and rained their calcium carbonate shells down onto the ocean floor.

Lovelock and the microbiologist Lynn Margulis had long argued that were it not for life, the Earth would have had a predominantly CO_2 atmosphere, not unlike that of our sister planets Mars and Venus.[6] This mostly CO_2 atmosphere, they argued, would have implied a greenhouse effect strong enough to have created searing earthly temperatures—some 60°C warmer than now.

Critics retorted that since phytoplankton hadn't even evolved for most of the time the Earth supported life, how could this mechanism to remove CO_2 have operated to solve the faint early sun paradox prior to the "recent" past (that is, the last couple of hundred million years)? One Gaian response speculated that algae, which have been in the oceans nearly since the beginning of life, can produce solid materials that could have captured some carbon. Indeed, the mushroomlike rocks we saw in our visit to the Archean shoreline were stromatolites, colonies of photosynthesizing blue-green algae living within hard carbon-containing materials that they secreted. Their living descendants can still be located today, most notably at Shark Bay in Western Australia.

Stromatolites go back fifty times farther into evolutionary history than the last of the dinosaurs. That such living structures could have existed in sufficient abundance to remove the requisite amounts of CO_2 over geological time, however, has not been quantitatively demonstrated; the issue remains a debate.

Sometimes radical new theories emerge that reconcile disputes among competing hypotheses—what the philosopher Thomas Kuhn referred to as conflicting paradigms. Although it is too early yet, in my opinion, to declare Gaia real, her advocates have advanced some clever new ideas supporting the biological mediation of climate change as a solution to the faint early sun paradox. For example, David Schwartzman at Howard University and Tyler Volk at New York University proposed a radical departure from the conventional wisdom (see figure 1.1) that Archean temperatures were neither very hot nor very cold. Instead, these scientists suggested that only when inorganic factors let the Earth's surface temperatures decrease sufficiently (to as "low" as 60–70°C, or 140–160°F) to allow primitive bacteria to survive did the Gaian control begin. Then, over the next several billion years of biological evolution, culminating with the appearance of trees and flowers a few hundred million years ago, the growing presence of life drew down the CO_2 and ended the super greenhouse effect. As the temperature low-ered over the eons, even more forms of life were able to sur-vive, and these, in turn, joined in the Gaian process of nega-tive feedback through CO_2 removal.[7]

Volk and Schwartzman suggest a specific mechanism, "biotic enhancement of weathering," for such CO_2 removal: Biota in the soils increase the contact area of minerals with weathering chemicals, thus greatly accelerating what inor-ganic processes could accomplish without life. The authors of this radical idea acknowledge one serious conflicting geo-logical fact: clear evidence dating from over two billion years ago showing episodes of polished and scratched rock surfaces, the same kinds of scouring that modern glaciers are observed to have made. Thus, conventional geological wis-dom asserts there were many glacial episodes (figure 1.1) over most of Earth history.[8] If so, then paleoclimatic history from the Archean until the evolution of complex organisms some 600 million years ago could not have been nearly as hot as this new Gaian theory requires. However, these Gaians, like good trial lawyers, struggle to produce alterna-tive explanations for evidence that contradicts their theories. In this case, Schwartzman and Volk suggest the rock scratches reveal debris flows following large impacts from extraterrestrial objects, like meteors and asteroids. A resolu-tion to this paradigm conflict is one of the rewards that pur-suing Earth systems science holds in store for the curious.

Despite the debate over CO_2 removal and paleotempera-tures, no one in the field doubts the essential role of life in producing that most critical substance in the air: oxygen. Photosynthesis uses solar energy to convert CO_2 and liquid water to carbohydrate and molecular oxygen. The reverse of this reaction is respiration and decay, in which carbohydrate and oxygen combine, liberate heat, and produce CO_2 and water vapor. It takes energy from the sun to break up an inor-ganic CO_2 molecule into organic carbohydrate and oxygen. Likewise, the process of respiration or decay uses oxygen

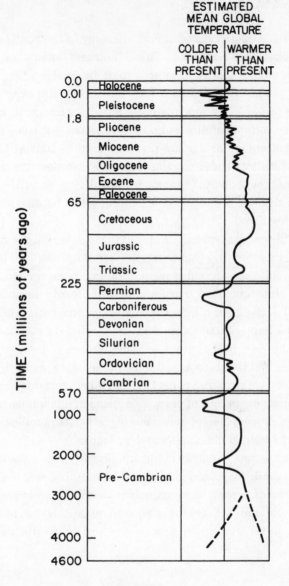

FIGURE 1.1

A schematic representation by L. A. Frakes of globally averaged surface temperatures over the indicated geological periods. Only relative departures from today's conditions should be inferred, and absolute amounts of change are quite uncertain. The graph represents the conventional view that, although most of paleoclimatic history has been warmer than today's climate, glacial eras have occurred from time to time over the past two billion years.

and releases that stored chemical energy in the bonds of the carbohydrate molecules. That "biomass" energy is what gives warm-blooded creatures their internal heating; it is why wood is burned as a fuel; and it explains why fossil fuels work: They are fossilized remains of organic matter whose carbon-based organic molecules contain some of the ancient energy of the sun that was used to convert CO_2 to plant matter. These remains somehow escaped the normal decay process that is the fate of most living materials—that is, such fuels contain carbon molecules that became trapped and fossilized.

This escape typically occurred through burial in environments devoid of oxygen, like the bottom of shallow inland seas, followed by subsequent trapping, compression, and, over time, chemical transformation of organic detritus into fossil fuels: coal, oil, and methane (natural gas). When we burn a lump of coal today, we're recovering the carbon dioxide and solar heat of dinosaur times trapped in fossil organic matter. While it took millions of years to make a coal deposit, we're releasing the CO_2 and other embedded chemical elements in tens of years. The speed of this human-accelerated process creates one of our biggest global problems and worries both climatologists and ecologists.

But before scientists could intelligently argue the issues I've raised, they needed ways of determining age, not only the age of a rock in terms of the date it was formed (the absolute age), but also the comparative age of layers, strata of rock, in relation to the layers above and below (the relative age).

DATING AN OLD PLANET

The Earth wasn't always thought to be billions of years old. In the 1700s, great debates took place in France and England about the dating of the Earth, at first between theologians and scientists and later among scientists themselves. Theologians were among the first to assign a number to the Earth's absolute age, and often it was heresy to question it. In 1654, using the Bible as a reference (he essentially counted "begats" backward), the Irish Archbishop of Armagh, James Ussher, announced that Creation had taken place on October 26, 4004 B.C., at 9:00 A.M. By the early nineteenth century, it had become obvious to most geologists that the archbishop's date could not possibly be correct—or even remotely close.

In the 1700s, the Scottish geologist James Hutton and some of his contemporaries believed that the physical and chemical processes that shaped the Earth's surface provided clear evidence that the planet was at least tens of millions of years old. They based this estimate on a geological principle known as uniformitarianism (as opposed to catastrophism). Uniformitarianism held that geological processes in the past were essentially the same as those at work today (a concept that will arise repeatedly in these pages). For instance, Hutton noted that the manner in which clay and silt were being deposited at the mouth of a river was observable. By studying new layers of sediments and how they solidified into shale and siltstone, he could infer how long it took the old layers to build up. By assuming that the same processes of sedimentation occurred over millions of years in the past, he could estimate the age of similar formations and hence arrive at the approximate age of the Earth. This method of using rock deposits as age indicators, however, depended on many variables that were too complex to measure. For example, climatic changes or differences in land elevation could alter the rates of both erosion and deposit.

Figure 1.1 shows the geological time sequence now established, based on modern dating techniques. It should be noted that in the nineteenth century, while geologists did not know the absolute age of the rocks they studied, they had an idea about the relative age of a stratum. The assumption that younger rocks developed over older formations is true most, but not all, of the time. This principle is called the Law of Superposition.

Geologists devised a system of nomenclature for the composition and age of various rock strata. This system consists of divisions of time and names for the rocks deposited during those times. The time units are simply subdivisions of geological time. The three major eras of "recent"—in the last 600 million years—geological time are the Paleozoic (570 to 225 millions of years ago—MYA), the Mesozoic (65 to 225 MYA), and the Cenozoic (present to 65 MYA), during which most life forms evolved. The fossilized remains of these forms have been used extensively to determine relative ages and to subdivide the eras into periods or epochs. We know many of the steps of evolution, which plants or animals came before other plants or animals, and thus we can determine the relative age of the rocks by which fossils they carry. (Fossils are valuable for another reason as well: By matching identical fossils from different places, it is possible to trace the motion of the continents across the surface of the Earth. Fossils are also, as we shall see, useful indicators of past climates.)

The nineteenth-century geologists established time sequences that largely solved the problem of relative dating, but absolute dating techniques were still sought. Finding a reliable clock to reconstruct the geological calendar became a major pursuit.

Lord Kelvin's Dating Method

Two hundred years after James Ussher proclaimed the Earth's age, Lord Kelvin tried to formulate another answer using scientific methods and reasoning.

This scientist, chair of natural philosophy at the Univeristy of Glasgow in the Victorian era, was one of the most influential theoretical physicists of his day. He used the known principles of thermodynamics to determine the age of the Earth. From observations of lava flows from the Earth's interior and from past explorations of deep mines, he knew that the interior of the Earth was hotter than its surface. By examining the temperature change between the Earth's surface and its interior, what is called the geothermal gradient, Lord Kelvin hoped to infer the Earth's age. He assumed that the planet began as a molten body, at about 3850°C (7000°F). His calculations suggested that the time needed for the geothermal gradient to reach its present value was close to 100 million years. That, he asserted, was the age of the Earth.

Kelvin's estimate set off a flurry of controversy among the theoretical physicists, who backed him, and the geomorphologists (geologists who look at land forms), who did not. The latter, using uniformitarianism to calculate how long certain geological features took to be created, believed that some features of the Earth's surface indicated the planet was much older than 100 million years. They could not, however, prove conclusively the greater age indicated by these observed geological features, so many physicists rejected the geologists' contentions. In this case, the new understanding the physicists presented actually increased scientists' uncertainty over the absolute age of the Earth. But only for some decades.

Sometimes new scientific insights or discoveries lead to a revised paradigm and reconciliations between seemingly conflicting theories and observations. Such was the case in this controversy: The question of the Earth's age became moot when radioactivity was discovered around 1900. The extra heat that radioactive processes generated in the Earth's core was unknown to Lord Kelvin, and if included in his calculations, would have brought his numbers closer to the geologists' best guess, in which they applied the principle of uniformitarianism to evolution of land forms. Radioactivity also provided an independent benchmark that geologists needed for absolute dating.

Radioactive Dating

A radioactive atom disintegrates over time. The disintegration rate of radioactive atoms is believed to be constant (unless the atom is traveling near the speed of light); it is virtually unchanged by pressure, temperature, or physical alterations of the compounds in which the atoms are contained (such as rock, water, or air). The decay rate of a radioactive element is expressed in terms of its half-life—that is, the time it takes for half the atoms originally present to decay into other (daughter) elements and particles by spontaneously emitting mass and energy. If we know at what constant rate daughter products are formed, all we need to determine a rock's age is the ratio of original element to daughter product. This knowledge enables scientists to calculate when the mineral containing the original radioactive element was formed. Using radioactive decay in rocks as a guide, geologists and geochemists can determine the absolute age of rocks, and hence of various strata of the Earth. With this technique, called radiometric dating, the elements for a reliable clock had been found.

Several radioactive elements are used to date rocks. Among them are uranium isotopes, which can decay (with half-lives between 700 million and 4.5 billion years) to isotopes of lead; rubidium, which decays (with a half-life of 50 billion years) to strontium; and potassium, which decays (with a half-life of 1.3 billion years) to argon.

Scientists experimenting during this early phase (from 1900 to 1938) of radiometric dating were hampered by crude analytical methods and an inadequate knowledge of the nuclear processes involved. But they were able to make

rough estimates by measuring lead-uranium ratios in uranium minerals and helium-uranium ratios in a variety of
rocks and minerals.

Because of long half-lives, the rubidium-strontium and
potassium-argon techniques are among the most reliable and
can span the entire history of the Earth—now put at about
4.5 billion years.[9] However, radioactive elements with
shorter half-lives are needed for dating more recent events,
such as those of the past few thousand years.

Radiocarbon Dating

In 1947, the American chemist Willard Libby discovered
an indispensable dating tool that enabled climatologists,
oceanographers, geologists, and archaeologists to reconstruct
accurately climatic change, geological events, and animal
and cultural evolution. Libby and his co-workers found a
way to estimate the age of the remains of plants and animals
that have died within the past forty thousand years or so,
such as wood and other plant remains (such as peat beds),
marine and freshwater shells, and groundwater and ocean
waters in which carbon has dissolved.

Carbon-14 (^{14}C) is one of the three naturally occurring
isotopes of the element carbon, which is abundant in the
atmosphere-ocean-biosphere reservoir. Unlike carbon-12
and carbon-13, which are stable carbon isotopes, ^{14}C is
unstable, but its atmospheric concentration is replenished
by a fairly continuous stream of cosmic rays impinging on
nitrogen molecules in the atmosphere and turning them into
^{14}C. (There are variations in ^{14}C production caused by variable solar activity, but these are not a major obstacle to the
use of carbon dating.) Through photosynthesis, atmospheric
carbon (including ^{14}C) is converted into organic carbon compounds.

While the plants are alive, there is a relative equilibrium

in the amount of ^{14}C in their tissues, since they are always replenishing their supply of ^{14}C by photosynthesizing atmospheric carbon. Because animals are always eating live or recently dead plants (or plant eaters), they also contain a ^{14}C level that is in close equilibrium with the atmosphere. But when the plant or animal dies, the radioactive ^{14}C in their tissues decays, since no further ^{14}C is taken up.

Like all radioactive elements, ^{14}C has a half-life. It is about 5,750 years: It takes approximately 5,750 years for half of a given number of ^{14}C atoms to undergo radioactive decay. Since this rate of decay is not influenced by outside conditions, the rate of disappearance of ^{14}C from a sample has an absolute relation to the time it was incorporated into the sample. Therefore, scientists can determine the age of a sample by measuring the relative amount of ^{14}C it contains.

Climatologists use ^{14}C dating along with the traditional technique of recording the strata of fossils to help establish a chronology of climatic changes. By using ^{14}C dating to determine the ages of samples from trees plowed under by ice masses, we can recreate the chronology of the continental glaciers' advance. The radiocarbon dating of peat samples from bogs and driftwood from lake shorelines has also yielded glaciation timetables. The ^{14}C content of shells of different planktonic animals found in deep-sea sediments enables us to date fluctuations in oceanic conditions conducive to each animal's abundance. In this way, we are able to infer temperatures and related climatic conditions. Climatologists have been able to use Libby's dating tool to obtain a worldwide picture of climate for about the past forty thousand years. Radiocarbon dates of charcoals from the hearths of cave dwellers have helped anthropologists sketch human history and its relation to climatic fluctuations.

ORIGIN OF OXYGEN
....

The burial of fossil carbon is critical to the story of how life influenced the atmosphere, not just because of the twenty-first-century prospect for the environmental risks of global warming from fossil fuel burning, but because that burial is part of the oxygen buildup story.

For a billion years or two, algae in the oceans produced oxygen. But since oxygen is highly reactive and since there was a great deal of reduced minerals (for example, iron that could easily be oxidized if it encountered oxygen) in the ancient oceans, most oxygen produced by life simply got used up before it could work its way into the atmosphere, where it also would be highly reactive. That is why most geochemists believe that for the first half of life's tenure on Earth, oxygen levels in the atmosphere were a very small fraction of what they are today. Even if evolutionary processes had "invented" more complicated life forms during this anaerobic era, had they tried to make a go at living on land or in the air, not only would they not have had oxygen to breathe, but the unfiltered ultraviolet light from the sun would likely have killed any such mutations before their evolution could proceed very far.

Geochemists have suggested that not until about two billion years ago, when most of the reduced minerals in the sea were used up, did oxygen begin to build up in significant amounts in the atmosphere. This then opened up an ecological niche for newly evolved life forms that need oxygen for energy to drive their aerobic metabolism.

Ozone to the Rescue

The presence of oxygen in the atmosphere had another major beneficial effect for a life form that might try to live at or above the surface: a filtering of biologically harmful ultraviolet radiation. UV breaks down many molecules, such as DNA, which is why there has been a virtual ban on the production of the infamous human-produced chlorofluorocarbons (CFCs) that are implicated in stratospheric ozone depletion. The two-oxygen-atom molecule, O_2, gets split by UV into the highly unstable atomic form O, which can combine back into O_2 and into a very special three-oxygen-atom molecule O_3, ozone. This molecule absorbs most solar ultraviolet radiation, and not until O_2 was abundant enough in the atmosphere could the atmosphere make the O_3 needed to allow life to get a (root- or foot-) hold on land. Indeed, it is not likely to be a coincidence that the rapid evolution of life from prokaryotes (single-celled with no nucleus) to eukaryotes (single-celled with nucleus) to Metazoa (multicelled) life all took place in roughly the last billion years or so—the era of atmospheric oxygen and ozone.

Volcano Climates and Drifting Continents

We should not get the impression that during the transition to oxygen and the removal of CO_2, the Earth experienced a uniform or uniformly changing climate (see figure 1.1). Neither climate nor atmospheric composition was static during the time of the explosion of biological evolution from single-celled bacteria or algae to Tyrannosaurus. Continents drifted and collided, mountains rose and eroded, volcanoes erupted, both catastrophically and continuously in midoceanic ridges, making the sea floor. The sea floor, being more dense than continental materials, sinks (subducts) underneath continental plates when they collide. One notorious "subduction zone" is the Pacific "Rim of Fire," a boundary between drifting continental plates where sliding and squeezing account for the high incidence of volcanoes and earthquakes at the rim: Kobe, Anchorage, San Francisco,

Los Angeles, Sakhalin Island, Mexico City. The materials squeezed under the continents into the lithosphere below the surface aren't necessarily lost forever from the atmosphere. Recall that some of those subducted materials were weathered rocks whose minerals combined with carbon removed from the atmospheric and oceanic system (a combination sometimes mediated by living processes) and became buried as sediments. Instead, extremely slow processes within the upper part of the "solid" Earth actually recycle some of these materials back into the atmosphere and oceans through continuous volcanic fissures or occasional explosive eruptions. This recycling could take hundreds of millions of years. The volcanic outgassing of recycled materials, the so-called sedimentary cycle, has further relevance for the CO_2 removal and climate stabilization story.

DRIFTING ALONG

Although there are many remaining uncertainties, the concept of continental drift was a new paradigm that revolutionized Earth sciences in the 1960s, when its basic principles became widely accepted by both geologists and geophysicists. Yet the original scientific proponent of the idea, the German meteorologist and geophysicist Alfred Wegener, first introduced it in the 1920s.

Continental drift had been a widespread, if fanciful, notion for centuries before Wegener. When maps of the Old and New Worlds were drawn, geographers noticed what a close, jigsawlike fit some of the continents, such as South America and Africa, would make if they could be moved together. Throughout the nineteenth century, geologists discovered similar rock and mineral layers, fossils, and other peculiar matches in corresponding places on different continents. For example, there are signs of the Permian glaciation in Southern Hemisphere continents such as South America and Africa, now an ocean apart, which led some to wonder whether these continents were once joined as part of one huge continent (dubbed Gondwanaland), positioned and glaciated near the South Pole.

The evidence includes similar patterns of rock grooves etched (presumably) by glaciers dragging boulders along the rock pavement. From these grooves, scientists have been able to determine the direction in which the paleoglaciers moved.

Skeptics scorned such proof and the theory the evidence supported. In the 1920s the president of the American Philosophical Society proclaimed continental drift to be "utter damned rot!"[10] And there the matter stood until after World War II. As it turned out, two more revolutionary developments were needed to give Wegener's ideas credibility: direct physical proof that the continents had moved (and that they are still moving) and an explanation of that movement.

Physical evidence for the movement of continents came with the discovery in the 1950s of a mid-ocean ridge system, some 65,000 kilometers of submarine mountain ranges that virtually circles the globe. This mountain chain was shown to have a narrow, deep rift along its center line. Up through that rift wells hot magma, which, as it solidifies, spreads outward and forms new crustal material. This process is called sea floor spreading, and by the 1960s it was verified by paleomagnetic techniques. As new rocks form, they become permanently magnetized, aligning their magnetic fields in the direction the Earth's field is pointing at that particular time. When the Earth's field reverses, these reversals show up on the sea floor in parallel stripes on both sides of ocean ridges. Oceanographers can determine just how fast the sea floor is spreading by measuring the distance from the rift to the magnetic reversals whose ages are approximately known.

The paleomagnetic evidence was gathered when magnetometers were dragged by research ships back and forth through the water over the sea floor "stripes" to determine their polarity. Direct evidence was obtained in the late 1960s when the Glomar Challenger extracted cores from the Mid-Atlantic Ridge as part of the Deep Sea Drilling Project. As predicted, the ocean floor proved youngest at the rift, and progressively older farther away on either side of the divide. Oceanic crust is continually being buried and recycled, which accounts for the young age of the sea floor relative to the age of the continents: The average age of the ocean floor is only about 100 million years, while the oldest terrestrial rocks have been dated at almost 4 billion years.

What happens to old crust when new crust is formed? And how does this renewal of crust affect the continents? The second revolutionary development that gave credibility to Wegener's theory—plate tectonics—resulted from attempts to answer these and other questions.

A distinction needs to be made between the observed facts supporting the existence of drifting continents and hypotheses to explain the causes of these facts. A similar distinction can be made in the example of biological evolution. As noted earlier, the fact of evolution is backed up by literally millions of bits of fossil and other, more modern evidence, even though the mechanism, Darwin's natural selection being the classical scientific theory, is still debated by scientists (and certainly challenged by creationists). That the theories explaining the process of these phenomena are not fully complete does not negate the vast amount of evidence supporting the existence of the phenomena themselves.

Similarly, regardless of how correct the widely believed theory of plate tectonics turns out to be, the evidence for continental drift is pervasive and is accepted by all knowledgeable geologists and geophysicists.

The notion of drifting plates was introduced by the late Canadian geophysicist J. Tuzo Wilson in 1965. According to the theory of plate tectonics, the Earth's crust is divided into huge segments, or plates, that travel on its pliable mantle, driven by heat from the Earth's interior. These plates, or continental and oceanic crust, are thousands of kilometers wide and up to 130 kilometers thick. When the plates move away horizontally from one another, they form fissures (the mid-ocean ridges) through which new crust can well up. When such fissures occur on land (one example is the Afar rift in Africa), continents begin to split. When plates collide, one of two things happens: They either buckle to form a mountain range or one subducts (dives under) the other, the process that produces the volcanically active Rim of Fire. Heat and pressure then melt some of the subducted material, which eventually reforms and upwells at fissures as new crust. When plates of continental crust collide, as they did when India rammed into Asia, mountain ranges are formed (in this case, the Himalayas). Earthquakes, as noted, also occur much more frequently at plate boundaries than elsewhere, which provides further support of the theory.

The important point for our purposes is that the surface of the Earth is constantly evolving. Earlier, we implied that continental drift was somehow connected to the era of more frequent ice ages. At least one climatic coincidence can be linked to the rearrangement of the continents: the isolation and subsequent glaciation of Antarctica.

Through much of the first half of the Tertiary Period, some 35 to 65 million years ago, before Antarctica was isolated, many species of deciduous (seasonal) and coniferous (evergreen) trees were present there. Fossil wood fragments on the West Antarctica islands attest to this fact. Hence, temperatures in Antarctica (at least at the margins of the continent, where the fossils have been found) were much warmer than today, by perhaps 10° to 15°C (18° to 27°F).

Fifty-five million years ago, Australia began to separate from Antarctica, which prefaced a marked climatic change on the latter continent and distinct biological evolution on the former—such as the isolated appearance of kangaroos. A circum-Antarctic ocean current has been unimpeded since the Drake Passage between South America and the Antarctic Peninsula opened up and a seaway was created between Australia-Tasmania and East Antarctica. This circular flow tends to isolate warmer waters to the north, which probably led to colder waters surrounding Antarctica. The development of Antarctic sea ice, perhaps 40 million years ago, and an ice cap later on are thus probably more than coincidental to the isolation of Antarctica. By four million to seven million years ago, Antarctica was certainly heavily glaciated, although some think it may have been ice-covered tens of millions of years earlier; the ice there today ties up enough water to coincide with about a 60-meter (190-foot) decline in sea level. There is evidence that another major current, the Gulf Stream, also grew over time.

While the currents were intensifying and the continents rearranging themselves to approximately their present positions, sharply differentiated climatic zones were being established between the equator and the poles. Individual plant and animal species that had once been widely distributed on a planet with warmer polar regions settled into more restricted regions with appropriate warm or cool climates, depending on their ecological requirements (which also change as species evolve biologically). While the ranges of many species became restricted, more ecological opportunities or niches were created, since there are more widespread climatic differences on today's Earth than in the early Cenozoic. This may be connected to the phenomenon reported by biologists like Harvard's Edward O. Wilson[11]: The absolute number of species (not necessarily their abundances) on Earth has increased as the planet cooled. Moreover, the high biological productivity that helped to create coal and oil beds so prevalent in the Cretaceous Period waned with the planetary cooling and possible reduction in CO_2 concentration in the transition to our era.

Thus the Earth eased into the Quaternary, the two-million-to-three-million-year recent geological period in which major ice expansions and contractions have recurred roughly every forty thousand to one hundred thousand years.

We are currently living through a ten-thousand-year-long climatically very stable interglacial epoch (the Holocene) within the Quaternary. How stable it will continue to be is addressed later.

....

CHAPTER 2

THE COEVOLUTION OF
CLIMATE AND LIFE

Climate influences and is influenced by life on Earth; the two appear to have coevolved. Both interact with wonderfully intricate cycles.

The environment is a complex network of cycles, all of which are critical to the beginning, evolution, and survival of life. Water constitutes rain and snow and the oceans, and leads to the deposit of sediments. Nitrogen, a critical nutrient, travels through its own texture of cycles, both in the atmosphere and into soils and water. Nitrogen is also linked to a sulfur cycle. Not only can sulfur produce acidic hazes and other potentially toxic conditions for plants and animals, but it plays a necessary part in the function of proteins. And that most important element for life on Earth, carbon, moves in a cycle linked to everything else. How these cycles work and what dangers, if any, lie within them are questions that can only be answered with some of the most modern instruments in the scientist's toolbox—in particular, satellites and computers. The development of simulation models for computers will be as close as most of us will ever get to that time machine.

Nutrients move in so-called biogeochemical cycles,[1] a term introduced in the 1920s by V. Vernadsky. It describes the interaction of life, air, sea, land, and other chemicals. One way climate makes its influence felt is by regulating the flow of materials through these cycles, in part through the vigor of the atmospheric circulation. In turn, the nutrients help determine the composition of the atmosphere, which then determines the climate. Water vapor is one such material. When it condenses to form clouds, more of the sun's rays are reflected back to space, thus altering the climate. Water vapor and clouds are also important elements in the greenhouse effect. But water is also one of the most important nutrients for sustaining life on Earth.

THE HYDROLOGICAL AND SEDIMENTARY CYCLES
....

At any one time, a vertical column extending throughout the depth of the atmosphere typically contains in vapor form half a million times less water than there is in the oceans and ice caps. The amount of immediately accessible fresh water falling over the globe each year as precipitation is also negligible compared to the water contained in the oceans. Yet this tiny fraction of all water that constitutes precipitating fresh water—which is continually distilled and distributed by the hydrological cycle—amounts to about 500,000 cubic kilometers of precipitation annually. This is enough to cover the 500 million square kilometers of the Earth's surface with about a meter of rainfall each year.

The energy source for the circulations of the atmosphere and oceans is, of course, the sun. It lifts the water from the oceans and lakes, and on land through evaporation. Later on, other factors such as condensation and droplet growth cause water to precipitate back to Earth. How water is distributed—in what quantities and which places—largely determines which life forms will thrive where.

Water is also transferred to the air from the leaves of plants in a process called *transpiration*. This, combined with evaporation from bodies of water and the soil, is known as *evapotranspiration*. Evaporation of ocean water is about six times larger in magnitude on a global average than evapotranspiration on land, although evapotranspiration can be the principal local source of water vapor in the centers of continents.

The precipitation that results from the hydrological cycle both creates and erodes the sediments. Water helps shuttle materials from land to sea, where they may ultimately end up as sediments. In the relatively shorter term, the sedimentary cycle includes the processes of erosion, nutrient transport, and sediment formation, for which water flows are mostly responsible. In the geologically longer term, the processes of sedimentation, uplift, sea floor spreading, and continental drift become important. Both the hydrological and sedimentary cycles are intertwined with the distribution of the amounts and flows of six important elements: hydrogen, carbon, oxygen, nitrogen, phosphorus, and sulfur, which are considered the macronutrients. These elements comprise more than 95 percent of all living organisms. Appropriate quantities of them in the right balance and the right places are required to sustain various forms of life. Although great stocks of all these nutrients exist in the Earth's crust in various (but not always accessible) forms, at any one time the natural supply of these vital elements is fairly constant. Hence, they must be recycled for life to regenerate continuously.

The Nitrogen Cycle

Nitrogen, an important nutrient, is also one of the most chemically complex, since it travels its cycle in many forms. Its primary form, nitrogen gas (N_2), makes up 78 percent of the atmosphere. Some of this gas is converted in the soils and waters to compounds containing ammonium nitrite, or nitrate groups. This conversion is known as nitrogen fixation, which describes what happens. Nitrogen is "fixed," or attached, to other chemical elements, and a strong chemical bond between the nitrogen and other atoms (typically hydrogen) is formed, a process also called nitrification. Nitrogen can be fixed abiologically by fires (including lightning or in car engines or in chemical fertilizer plants), or biologically by special nitrogen-fixing organisms.

Fixed nitrogen resides in the air, soil, and water. Special bacteria take energy from plants to do their work, fixing nitrogen. They often live in nodules on the roots of legumes, members of the pea family such as alfalfa, beans, peas, and clover. Because these plants are able to fix nitrogen, they are often planted between crop seasons to replenish the soil nutrient supply depleted by harvesting non-nitrogen-fixing plants such as wheat, corn, and tomatoes. This natural fertilizer allows plants to incorporate appropriate forms of fixed nitrogen into their tissues by absorbing it in their roots. The plants then chemically transform it into amino acids and convert it into proteins.

Nitrogen, fixed as proteins, for example, into the bodies of living things, eventually returns via the nitrogen cycle to its original form of nitrogen gas in the air. This process starts when the plants containing the fixed nitrogen either are eaten or die. If they are eaten, most fixed nitrogen is returned to the environment as animal excretion or carcasses. These fixed nitrogen products (including dead, uneaten plants) encounter decomposers such as denitrifying bacteria that can undo the work done by the nitrogen-fixing bacteria. When the waste products are denitrified, their fixed nitrate is transformed in several steps back into nitrogen gas, for the most part, but also into lesser amounts of nitrous oxide (popularly known as laughing gas).

Like water vapor and CO_2, nitrous oxide (N_2O) is a

"greenhouse gas" that can trap heat near the Earth's surface. Over many years, the nitrous oxide is transported by winds high into the atmosphere, where it is broken down by ultraviolet light. When nitrous oxide is destroyed by this process, other nitrogen oxide gases (NO and NO_2) are created. Interestingly, NO and NO_2 in the stratosphere are believed to help limit the amount of ozone. Nitrogen oxides in the atmosphere are chemically transformed back to either nitrogen or nitrate or nitrite compounds, the latter of which may get used by plants after they are washed by the rain back to the Earth's surface.

The Sulfur Cycle

Another example of a major biogeochemical cycle of significance to climate and life is the sulfur cycle. The nutrient sulfur plays an important part in the structure and function of proteins, thus influencing all life. While certain quantities and forms of sulfur can be toxic to plants or animals, others determine the acidity of rainwater, surface water, and soil. This acidity controls the rates of processes such as denitrification.

Like nitrogen, sulfur can exist in many forms: as the gases sulfur dioxide (SO_2) or hydrogen sulfide, or as the compound sulfurous acid, which, when exposed to sunlight, can change into caustic sulfuric acid. When sulfuric acid particles float in the air, they contribute to the irritating smog that engulfs some industrial areas where many sulfur-containing fuels are burned.

The sulfur cycle can be thought of as beginning with the gas sulfur dioxide or the particles of sulfate compounds in the air. These compounds either fall out or are rained out of the atmosphere, contributing to the sulfur compounds in the surface environment. Some forms of sulfur are taken up by plants and incorporated into their tissues. Then, as with nitrogen, these organic sulfur compounds are returned to land or water after the plants die or are consumed. Bacteria are important here too, since they can transform the organic sulfur to hydrogen sulfide gas. Some phytoplankton in the oceans can produce a chemical that transforms to sulfur dioxide that resides in the atmosphere. These gases can reenter the atmosphere, water, and soils, and continue the cycle.

The sulfur cycles generally work quickly, but other processes, including erosion, sedimentation, and uplift of rocks containing sulfur, take a long time. Sulfur is added to the environment by volcanoes and human activities, usually from industry. When we burn sulfur-containing fossil fuels, we release sulfur dioxide, which can mix with moisture in the atmosphere and contribute to environmental degradation in the form of acid rain. The sulfuric acid droplets of smog form a haze layer (called sulfate aerosol) that can cause lung disease and also modifies the atmospheric albedo (reflectivity), and can thus alter the amount of solar energy absorbed by the climate system—usually cooling the surface. Such sulfate aerosols, whether generated by industrial activities, phytoplankton, or volcanoes, can alter the brightness of clouds and the atmosphere, affecting the climate.[2] While many questions remain open, the sulfur cycle in general—and human-induced sulfate aerosols, acid rain, and smog issues, in particular—are major physical, biological, health, and social problems.

The Carbon Cycle

The cycle considered to be of greatest interest to global change is the carbon cycle. Carbon, we know, exists in trace amounts (0.035 percent currently) in the atmosphere as carbon dioxide (CO_2), and in this and other forms in larger amounts in the oceans, sediments, and rocks. Plants are able to use carbon to form carbohydrates and sugars that are used to build their tissues. They use solar energy to combine CO_2 and water in photosynthesis. The carbon dioxide uptake speeds up during the spring and summer, when increasing sunlight and warmer temperatures help plants take CO_2 out of the air at a faster rate.

Every year in the Northern Hemisphere, the concentration of CO_2 in the air drops by about 3 percent between spring and fall. This annual inhalation of carbon involves tens of billions of tons of CO_2. In the Southern Hemisphere, where there are fewer plants, the exchange of CO_2 between the air and vegetation is only about one-third that in the Northern Hemisphere.

With the onset of fall and winter, temperatures drop and photosynthesis rates slow, since less solar energy is available to convert CO_2 to carbohydrates. Then the other part of the carbon cycle in plants dominates, as respiration of living plants and the decay of dying plants or dead organic matter proceed at a faster rate than photosynthesis.

Of course, factors other than CO_2 are involved in the carbon cycle. CO_2 exchanges between air and the oceans are controlled by complex internal biological and chemical processes in sea water.[3] The location and quantity of plant life on Earth is another such factor. And, as we have seen, other nutrients such as water and nitrogen are required to sustain life. They interact with carbon and life in an interlocking set of biogeochemical cycles.

We mentioned that CO_2 is a trace gas in the Earth's atmosphere, which means, relatively speaking, that there is not much of it—currently 0.035 percent of the air—but the 750 billion tons of atmospheric carbon this small percentage represents has a substantial effect on the atmospheric heat balance. The climatic power of CO_2 lies in the fact that it tends to transmit most of the solar radiation but to absorb a much larger fraction of infrared radiation, trapping some of the Earth's radiant heat that otherwise would escape through the atmosphere to space (in other words, it is, as noted earlier, a "greenhouse gas").

There are other trace gases in the atmosphere with strong greenhouse effects that could increase in concentrations. Notable among these is methane. The concentration of CH_4 has gone up by about 150 percent since the Industrial Revolution; it is produced by animals and bacteria and as a pollutant from human activities such as mining and agriculture. Nitrous oxide also is increasing, perhaps as a result of the rapid growth in the use of nitrogen fertilizers.

The primordial CO_2 concentrations resulted from a combination of volcanic eruptions spewing the gas into the atmosphere, the formation and weathering of rocks, the synthesis and decay of organic matter, and the chemical transformation of undecayed organic matter into fossil fuels—all of which took place over the eons. We humans are digging up those fossil fuels and releasing them at a much faster rate than they were made. Satisfying our energy and agricultural demands has contributed a 20 to 30 percent increase of airborne CO_2 in the 150 years since the Industrial Revolution. Most projections point to a 100 percent increase by the middle of the twenty-first century as not unlikely.

THE GREAT PLAINS OCEAN
....

There has always been enough CO_2 in the Earth's atmosphere to support photosynthesis. We've also seen that CO_2 has constantly been removed by weathering, both through organic and inorganic mechanisms. If weathering had been too successful, there wouldn't be enough CO_2 left to support plant life—and we know that didn't happen. Here is where volcanic action, in particular, the continuous activity occurring under the sea at mid-ocean ridges, plays a role.

Melting all the ice grounded on Greenland would raise

worldwide sea levels only about 5 meters, and melting all the mountain glaciers of the world would add only a fraction more. Melting the vast glaciers on Antarctica would have a much bigger impact, but even this dramatic, unimaginable event would raise sea levels only on the order of 60 meters. This is still less than a quarter of the sea level rise that geologists tell us marked the high-water point in the Cretaceous Period—about a hundred million years ago, during the reign of *Tyrannosaurus rex*. Thus, ice melting, while part of the story, clearly wouldn't come close to raising sea levels sufficiently to cause the extensive inland seas of the Cretaceous or to have had only 20 percent of that ancient world covered by land rather than the 30 percent of today. What else might explain such high sea levels?

Two logical possibilities are that at that time there was more water on Earth, or that the continents were sunk deeper into the Earth's crust. These speculative ideas have no supporting evidence and are considered extremely unlikely by most scientists.

The most plausible explanation is relatively straightforward, once we allow ourselves to think in geological time frames: The volume of the ocean basins was less at this time in the early history of the Earth, and thus ocean waters flooded higher on the land than they would today. What could fill up the early ocean basins? Volcanic rubble from mid-ocean ridges, most likely, as these could have been manufactured a hundred million years ago more rapidly than in recent times. But if there were more undersea volcanism building up mid-ocean ridges, then there should have been more CO_2 injected into the system, since that is one of the substances outgassed through volcanism.

Although we have no direct means to measure atmospheric composition at the time of the dinosaurs, we do know several important factors: (1) it was very warm, about 10°C (18°F) warmer in mid-Cretaceous times than at present, which could have meant an enhanced greenhouse effect and rising atmospheric temperatures from more CO_2; (2) broadleaf vegetation was widespread throughout the Earth, and more CO_2 encourages increased photosynthetic activity; and (3) fossil fuels were then created in significant quantities. Fossil fuels being derived from buried organic matter suggests the possibility of higher plant or plankton productivity, once again plausibly associated with CO_2-enhanced photosynthesis. All this is only circumstantial evidence, of course.

Eric Barron of Penn State University has produced continental drift maps for the Cretaceous Period. Figure 2.1 shows that the world of a hundred million years ago was very different from today's geography.[4] Shallow mid-continental seas divided the western from the eastern United States. That would explain the existence of fossilized clams in the now-mile-high foothills of the Rockies.

Water absorbs more solar energy than land because it is darker on average, and so a planet with one-third less land would be a little bit warmer because of this effect alone. In addition, a planet that had no permanent polar ice, which nearly all evidence suggests applies to the mid-Cretaceous, would reflect less sunlight back into space than does the current world, with its white polar caps, and this, too, would contribute to warming. Factors such as these were fed into three-dimensional computer models of the Earth's climate system in order to estimate just how warm the mid-Cretaceous may have been.

One modeling study showed that the combination of changes in geography and absence of polar ice was sufficient to raise the Earth's temperatures some 5°C relative to today's. But 5°C is less than the warming that most paleoclimatic evidence suggests for the mid-Cretaceous. That model, produced by Warren Washington and Barron at the National Center for Atmospheric Research (NCAR) in Boulder, Colorado, was simply too cold to match the other evidence. In addition, because we know from the fossil evidence that broadleaf plants and alligators lived near the Arctic Circle,

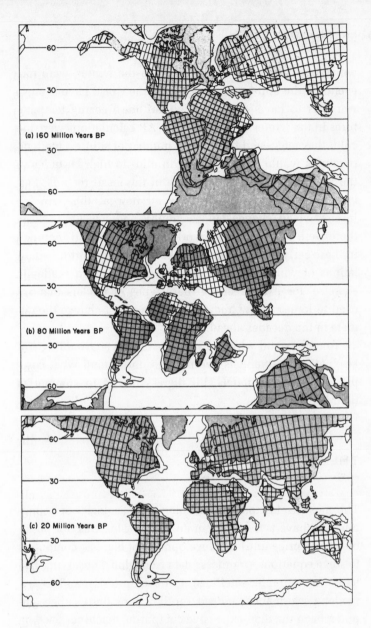

FIGURE 2.1

Three "snapshots" of paleocontinental positions reconstructed from data on continental drift by Eric Barron and colleagues. The shaded areas were above sea level at the times of each reconstruction. While detailed regional features should not be considered definitive, the broad scale patterns shown are well established. Present latitude and longitude lines are drawn on the continents to indicate how the land masses have drifted over time.

we assume that deep freezes, even in the winter, were rare events in the mid-Cretaceous. Could the world have been so warm as to have very few high-latitude freezing temperatures in the winter? Because the NCAR calculations were too cold, they suggested that a large number of subfreezing temperatures would have occurred in mid- to high latitudes in the wintertime, contrary to some of the fossil records. Perhaps imperfect models—we are all new at this—are too insensitive to changes in geography and ice. Or perhaps there were other factors at work as well, which, if included in these early models, would make them more faithful simulations of what happened. This is more than an academic exercise, for it is precisely these types of models that are used to forecast how human activities might change the climate in the decades ahead.

To understand why this kind of study is so troubling and fascinating to scientists, we need to understand what comprises computer models. I'll digress briefly to summarize their bare essentials.

THE ART OF MODELING
....

The most useful tool a climatologist or ecologist or economist can have is a fast, accurate model. This was not possible at any price until the development of big, fast computers to solve equations, to process data from global observing systems (for example, satellites), to develop ideas, and to test model performance. Indeed, until the modern supercomputer, even the then-expensive computing machines used by universities and big corporations in the 1960s were simply too slow to make enough calculations.

The father of computer modeling of the atmosphere was Lewis F. Richardson, a visionary greatly admired by scientists. Richardson tried to calculate the weather using mathematics in the 1920s, forty years before the widespread deployment of even the early generations of electronic computers. Weather forecasting had been practiced in London before Richardson, using Colonel E. Gold's *Index of Weather Maps*. Observing stations would telegraph current conditions to the Meteorological Office and were placed upon a large-scale map. The index then enabled the forecaster to find a number of previous maps that resembled the newly drawn one, and a forecast could be made based on the idea that what happened in the past would repeat. The history of the atmosphere acted as a "working model of its present self,"[5] a pattern-matching version of the geologists' uniformitarianism principle. Richardson sought a new paradigm for weather forecasting—a mathematical model of basic physical laws rather than mapping analogs.

The problem with analog maps, Richardson noted, was that the weather didn't always follow the same pattern. Although what has happened can happen again, it is not safe to assume that what will happen has necessarily already happened. Unique events and situations arise. So he proposed a scheme of weather prediction based on first principles in the form of differential equations, the mathematical expressions of known natural laws. Since the differential equations could not be solved exactly, he suggested using an approximation technique known as a numerical method. He produced a series of forms to help put the observational data into numerically convenient terms. Richardson fully understood that the computational power to actually predict the weather with his numerical techniques was only a dream. In his dream he envisioned a huge facility, "a large hall like a theatre,"[6] housing hundreds of human "computers" who would calculate weather from these forms. Using the numerical rules—what today are called algorithms—obtained from the original differential equations, his preliminary attempts all failed, but not because the basic idea was wrong. Rather, Richardson simply did not realize that the approximation techniques he chose would give nonsense answers unless applied slightly differently than he did. Decades later, and with the funding that the nuclear arms race engendered,[7] mathematicians discovered how to make Richardson's numerical methods work. In fact, they are the basis of weather and climate models routinely used today.

The advantage of modeling is that it enables us to conduct experiments impossible or impractical in the real world. Essentially, a model is a series of mathematical equations, coded into computer algorithms, that recreate in a computer something that models reality. It enables the scientist to pose a series of what-if questions—in a way, to play harmlessly with nature on a grand scale: If one thing changes, what happens to everything else in the climate? If I change one variable, say, the radiant power of the sun, what will happen to variables such as temperature and rainfall? And finally, since models are rarely perfect replicas of reality, how much of their answers should we believe?

To build a model of any system, one must decide beforehand what components of the system to include. To build a model railroad, for instance, you must include the basic components, such as tracks, and then choose which cars to replicate. There are other features to consider, depending on how realistic a replica the model railroad is to be: for example, water towers, switches, signals, stations.

To simulate the climate, a modeler needs to decide which components of the climatic system to include and which variables to involve.[8] For example, if we choose to simulate the long-term sequence of glacials and interglacials, our model needs to include explicitly the effects of all the important interacting components of the climatic system operating over the past million years or so. As we have seen, life influences the climate, and thus must be included too. These mutually interacting subsystems form part of the *internal* components of the model.

On the other hand, if we are interested only in modeling

very short-term weather events—say, over a single week— then our model can ignore any changes in the glaciers, deep oceans, land shapes, and forests, since they change very little in that time span. Those factors would be called *external* to the modeled climate systems.

Modelers speak of a hierarchy of models that ranges from simple Earth-averaged, time-independent, temperature models (in other words, models that give you the average temperature for the entire planet over a wide time span) up to high-resolution, three-dimensional, time-dependent models. The latter would include variables in the atmosphere, oceans, biosphere, and sometimes even the Earth's crust. As you can imagine, these more comprehensive models also are extremely complex, difficult to build, expensive to run, and tricky to test. One would hope that the added complexity would add reality to the simulation, but this is not always the case, making the art of modeling an often contentious enterprise.

After we decide the processes and subsystems we want to factor into our model, we write the algorithms that best describe those variables in such a way that a computer can execute our commands. We accept, sometimes on faith, that variables in the climatic system interact in accordance with the laws of nature as we understand them and have written them down mathematically. The degree of resolution and physical comprehensiveness of our model determines how many and what kind of expressions must be written down in order to make our particular model a reasonable approximation (we hope!) of the known laws. For very simple models, the mathematical equations that describe the behavior of the climatic variables can be solved analytically by any high school freshman who knows elementary algebra. However, as soon as the climatologist tries to include many climatic variables or more than one space dimension in a model, the complexity of the mathematics and the computer algorithm it spawns increases drastically. To compute a few days of weather at each of about forty thousand patches around the world typically takes an hour of computer time on one of those electronic behemoths known today as supercomputers.

The original equations for weather or climate models typically express the value of each climatic variable continuously over space and time. But the actual equations solved in the computer are approximations of the original. Consider temperature. Instead of solving an equation exactly for temperature everywhere, modern computers use approximation techniques, which include data from a network or grid—discrete points in space and at a discrete time. Anything that happens between those spaces on the grid, or at any time other than the one being measured or computed, needs to be averaged in. The newest techniques approximate better what happens between the grid points. Local phenomena, such as lakes or mountain valleys or individual thunderstorms, can alter local conditions, and that would not be seen in the computer code if the grid squares are very coarse—they are typically hundreds of kilometers across in today's practice. (Techniques to deal with such "subgrid scale" phenomena are described in more depth in chapter 4.) The only other solution to the problem would be more grid points, which means more computation and more needed data—a very expensive process. Costs go up ten times every time the size of the grid is cut in half.

MODELING THE DINOSAURS' CLIMATE

Back to that problem of the Cretaceous temperatures being too cold in the NCAR model. Several of us in Boulder in 1984 pursued the solutions to this problem using a complex computer model.[9] We tried various combinations of assumed Cretaceous ocean temperature patterns in our model to see if we could find a way that ocean currents might have prevented the mid-continental, midwinter freezes in high latitudes that the model predicted. We even went so far as to assume the North Pole oceans had the same warm surface temperature as the rest of the world. In none of the simulations, however, could we stop the long winter nights from radiating enough infrared heat radiation into space to cause severe episodes of inland frost—at least when, by chance, winds were not blowing from the warm oceans over the dark, high-latitude wintertime continents. One possibility emerges from the faint early sun paradox debate: an enhanced greenhouse effect from extra CO_2 in the atmosphere. But how or where did this extra CO_2 come from?

Also in the 1980s, some of our colleagues at other laboratories, most notably Robert Berner at Yale University, noted that geological evidence, which suggests varying sea floor spreading rates, showed clear signs that the mid-Cretaceous, a hundred million years ago, was a period of very active undersea volcanism and sea floor spreading.[10] This helps tie together the story, since that would both pile up volcanic rock under the oceans at fast rates, thereby diminishing ocean basin volume and raising sea levels, and inject vast amounts of CO_2 into the ocean-atmosphere systems. They proposed a combined organic and inorganic feedback mechanism that was built on top of the Gaian and WHAK mechanisms discussed earlier. When sea floor spreading rates were high, sea levels were high, CO_2 was high, and climate was warm and therefore wetter. Warm and wet climate with high CO_2 levels, they suggested, increased the weathering and phytoplankton production rates, which in turn, removed some of the abundant CO_2 by inorganic weathering and biological burial as carbonate sediments.

This, then, provided a negative (or stabilizing) feedback to remove CO_2, thereby preventing the climate from becoming overdramatically warm. In other words, on "short" time scales of tens of millions of years (rather than hundreds of millions to a billion years), factors such as varying rates of continental drift, volcanic activity, and biological activity could combine to vary carbon dioxide concentrations in the air by perhaps as much as five times relative to the present. Berner and his colleagues' model predicted CO_2 levels several times greater than the present levels around the mid-Cretaceous.

In the absence of clear direct evidence, this entire picture is consistent but circumstantial—a paleoclimatologist's bedtime story, if you will—rather than proved beyond reasonable doubt.[11] That is why scientific debates are set off when we rely too heavily on untested computer modeling. Unfortunately, no other tools can perform what-if experiments. The art is to ask them questions they can credibly answer. It is no simple skill.

Geochemists often support the notion that CO_2 concentrations in the atmosphere decreased, along with a decrease in sea floor spreading rates, over the 100-million-year transition from the Mesozoic to the present. The latter began with the end of the Cretaceous and the extinction of dinosaurs and half the other species then alive, about 66 million years ago.

Much has been written about the cause of the end of these wondrous creatures, and explanations for the dinosaur demise range from biological competition, disease, and other "internal" causes, to a massive collision between the Earth and an asteroid or a comet 10 kilometers or so across. The explosion from such a violent crash might have spewed enough material into the atmosphere to obscure sunlight for months, perhaps years, depriving the Earth of photosynthesis, causing freezing temperatures on land (so-called asteroid winter), and making enough nitric acid in the atmosphere from the shock waves of the collision to acidify the oceans. All of these impacts could also have temporarily wiped out the ozone layer and dramatically altered the greenhouse properties of the atmosphere. That such an external catastrophe could have created a synergism of deadly disturbance that destroyed the remaining dinosaurs and half the other species extant at that time is plausible and even widely accepted, although debate still rages over the details. Its credibility rose considerably in the early 1990s when geologists found what appears to be the imprint of the impact crater in the Yucatán peninsula.

Catastrophic events notwithstanding, it would not be strictly correct to say that the climate has been cooling steadily for the last hundred million years, since there were periods of relative warmth or coolness, some lasting for millions of years, during the transition to modern geological time (see figure 1.1).

Nevertheless, the chemical composition of fossil plankton shells whose occupants once lived near the ocean bottom provides a measure of bottom water temperatures throughout this part of geological history. Over the past hundred million years, bottom waters appears to have cooled by up to 15°C toward their current global average of around 0°C. Sea levels dropped by hundreds of meters, and continents drifted apart to roughly their current positions. Inland continental seas largely disappeared, with a few remnants like the Persian Gulf, and the surface climate cooled some 10°C on average. Somewhere around fifteen million to twenty million years ago, with the opening of the Drake Passage between Antarctica and South America, permanent ice appears to have built up on Antarctica. Some speculate, as already noted, that the circumpolar Antarctic current that was then allowed to develop because of this geographical change in continental positions and ocean floor shape isolated the Antarctic continent from the warm ocean currents that previously may have touched its shores, thereby preventing continental-scale permanent ice, as now exists. Others speculate that declining atmospheric CO_2 concentrations during the Cenozoic allowed ice to gradually build up on the south polar continent.

A TILTING EARTH
....

About two million to three million years ago, the permanent ice coverage of the Arctic Sea appears to have been established, and the paleoclimatic record starts to show significant expansions and contractions of warm and cold periods separated into approximately forty-thousand-year cycles. This is an interesting number, since it corresponds roughly to the time it takes the Earth's pole to complete an oscillation from about 22½ to 24½ degrees of tilt from the plane of the Earth's orbit. Currently, the axis is inclined at 23½ degrees, which accounts for the latitudes of the current Tropic of Cancer and Tropic of Capricorn. The maps of a few thousand years from now will need to redraw these celebrated demarcation lines tens of kilometers closer to the equator. Physically, that means less tilt of the axis will have less contrast in the heating between winter and summer by a few percent.

It has long been speculated, and recently calculated, that such a change in the amount of sunlight coming in between winter and summer seasons—especially at high northern latitudes where big ice sheets can grow—could be responsible for initiating or ending ice ages. This is known as the Milankovitch mechanism. Most interesting, and perplexing, perhaps, is the discovery that sometime about six hundred thousand to eight hundred thousand years ago, the principal cycle in cold and warm extreme episodes became an intense hundred-thousand-year difference between the peaks of the interglacials (although the forty-thousand-year cycle remained as a weaker beat). The last major phase of glaciation ended about ten thousand years ago. At its height, twenty thousand years ago, mile-high ice sheets covered much of northern Europe and North America (figure 2.2).[12] Glaciers expanded in the high plateaus and mountains all around the world. Tropical forests contracted, and deserts generally expanded.

Enough ocean water was locked up on land as extra glacial ice to have dropped sea levels by over 100 meters relative to today. The massive ice sheets scoured the land and shaped the terrain. The ice age temperatures, globally averaging around 5 to 7°C cooler, revamped the ecological face of the Earth.

Why did the hundred-thousand-year cycle come to dominate the shorter and weaker ice age interglacial cycles some eight hundred thousand years ago, and why did the cool/warm cycles develop at all several million years ago? Definitive answers remain to be found, although some ideas and calculations are quite promising.[13]

ANCIENT AIRS
....

One of the most interesting discoveries in the past twenty years in Earth sciences has come from ice cores drilled in Greenland and Antarctica. When snow falls on these frozen continents, the air between the snow grains eventually is trapped as air bubbles as the snow is compressed into ice, and some of these bubbles are two hundred thousand years old.

Scientists are bringing up to the surface hundreds of 5-meter-long sections of ice from as deep as 2,000 to 3,000 meters, then flying this frozen cargo by ski plane to the coast for the long trip back to laboratories in France, Switzerland, Denmark, and the United States for analysis of the chemical contents of the ice and air bubbles. These cores offer us a library of the history of the Earth's atmosphere, stretching back to our Neanderthal ancestors one hundred thousand to two hundred thousand years ago.

In the labs, thin slices of ice are melted in sealed chambers and the trapped gas bubbles are freed for detection by sensitive instruments. From such work we know, for example, that air breathed by the ancient Egyptians and Anasazi Indians was roughly speaking very similar to the air we breathe, except for a whole host of air pollutants introduced in the past hundred or two hundred years. Principal among these pollutants are extra sulfur dioxide, extra CO_2, and extra methane. CO_2 has, I've noted, increased by 25 percent due to industrialization and deforestation, and methane has increased 150 percent from a variety of human activities associated with agriculture, land use, and energy production. Natural changes are also detected, such as the acid snows from big volcanic eruptions.

To me, the most remarkable finding from the ice cores is

RETREAT OF ICE (thousands of years ago)

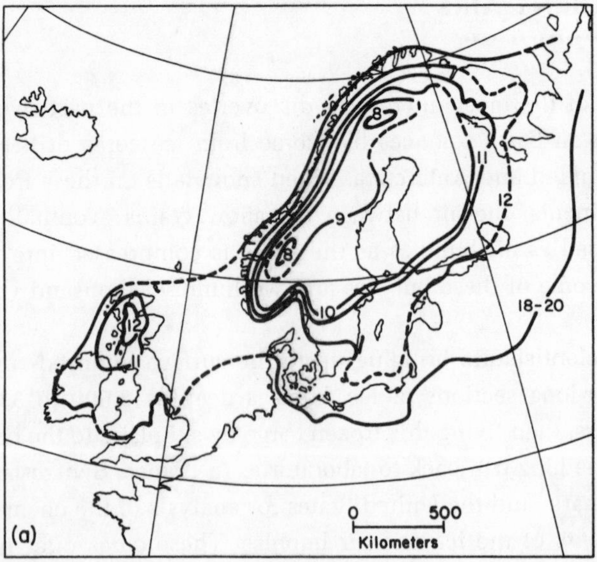

FIGURE 2.2

Where the ice edge was (in thousands of years of age) is indicated by the number on the contour lines. As the last ice age ended and the (a) Fenno-Scandinavian and (b) Laurentide ice sheets melted, climatic conditions generally moved toward their interglacial states. By examining the earliest dates of remains suggesting that the ice sheets had gone, scientists have reconstructed maps such as those shown here. However, it would be misleading to think not the relative stability of the global climate and the greenhouse gases in the atmosphere over the ten thousand years of human civilization. Rather, the ice at both poles shows that during the height of the last ice age there was, on average, some 30 to 40 percent less CO_2 and half as much methane in the air as there was during most of the Holocene, our time. (Actually, this is true only for the part of the Holocene preceding the Industrial Revolution and its pollution.) And the same direct relationship between less/more greenhouse gases and lower/higher temperatures has been found for the ice age peak and interglacial before that, some 120,000 to 150,000 years ago (see figure 2.3).[14]

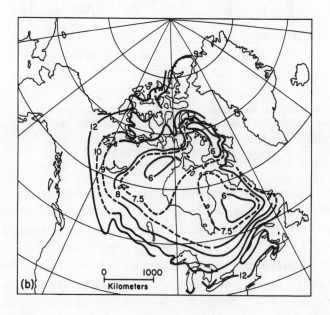

that the ice sheets simply melted in place following the progression of time lines seen here. Large fractions of the Pleistocene ice sheets, particularly the North American one, could well have melted in the oceans after having been discharged more than 13,000 years ago as icebergs into the sea. Thus, these maps could show simply when the final remnants of snow and ice cover melted on land.

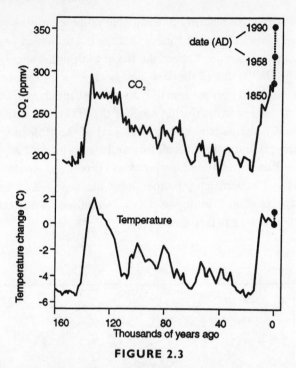

FIGURE 2.3

Data from Antarctica show that CO_2 and temperatures are highly correlated over the past 160,000 years. On this geological time scale, the CO_2 and temperature changes since preindustrial times appear as sharp spikes relative to the slower rates of natural change inferred from ice-core data near the South Pole.

This remarkable finding suggests the possibility of a positive (destabilizing) feedback rather than a negative feedback among CO_2, methane, and climate change. That is, when the world was colder there was less of these greenhouse gases, therefore less heat trapping and enhanced cold. As the Earth warmed up, CO_2 and methane increased, accelerating the warming. The Gaia hypothesis states that life should act to control environmental conditions to preserve climatic stability. Yet if life had a hand in this temperature-CO_2-methane story (which is likely), it would have been accelerating, rather than decelerating, the change of climatic conditions. Once again, the scientific story is still incomplete. Nevertheless, most scientists would agree that life has been a principal factor in the positive feedback between climate change and greenhouse gases, between recent ice ages and interglacials.

What is important for our discussion isn't the absence of high confidence over which particular mechanism explains the role of life in the CO_2-climate positive feedback, but the fact that the feedback appears positive. This is different from what we inferred from studying the one- to two-billion-year-long transition from the high-CO_2, low-oxygen atmosphere of the Archean to the era of great biological evolution about half a billion years ago. For that transition, as noted earlier, life may have been an important contributor to the CO_2 removal process that helped stabilize climate—the negative feedback implicit in the Gaia hypothesis.

In this connection, the geochemist Robert Berner revised his earlier geochemical model to include half a dozen improvements and new factors, ranging from new data on inorganic factors such as sea floor spreading and long-term solar luminosity increases to organic processes resulting from the evolution and spread of vascular land plants some 300 million years ago. The latter, Gaian scientists have said, resulted in accelerating the chemical weathering process in the soils.

"As a result," Berner notes, referring to the widely believed cold period about 300 million years ago (see figure 1.1) at the Permian-Carboniferous transition period,

the removal of atmospheric CO_2 by the massive burial of sedimentary organic matter during the Carboniferous and Permian periods, evidenced by the abundance of coal of this age, amplified the large mid-Paleozoic drop in CO_2 concentration. The consequently lowered greenhouse effect from the CO_2 drop likely had a major influence in bringing about the Permian-Carboniferous glaciation, the most extensive and longest glaciation of the entire

Phanerozoic [past 570 million years]. These results further support the idea that the atmospheric greenhouse effect has been a major factor affecting global climate change over geologic time.[15]

Aside from the extra confidence in our understanding of the greenhouse effect that Berner's work contributes, it also is relevant to the organic-inorganic climatological debate. He wisely counsels open-mindedness: "a purely geologic or purely biologic approach to the long-term carbon cycle is overly simplistic."

IS GLOBAL CHANGE UNPRECEDENTED?
....

The greenhouse effect theory of heat trapping, codified in mathematical models of the climate, suggests that when CO_2 (or its heat-trapping equivalent in other greenhouse gases, like CH_4) doubles (sometime in the middle of the next century, should population, economic, and technology trends continue as typically projected), then the world will warm up somewhere between 1° and 5°C by A.D. 2100.

Even the mild end of that range would mean warming at the rate of 1° per hundred years, ten times faster than what has been the average rate of natural sustained global temperature change from the end of the last ice age to the present interglacial. Should the higher end of the range occur, then we could see rates of climate change some fifty times faster than sustained, natural average conditions. Global climate change at this rate would almost certainly force many species to attempt to move their ranges in a struggle to keep up with rapidly changing climatic conditions, just as they did from the ice age to interglacial transition, ten thousand to fifteen thousand years ago.

What appears unprecedented is the combination, or synergism, of potentially very rapid rates of human-induced change at the same time that nature has been fragmented and assaulted with a host of chemical agents or transplanted "exotic" species that do not naturally occur where they are introduced by humans. It is for these reasons that it is essential to understand whether the projected doubling of CO_2 will warm the Earth by 1° or 5°, for that is the difference between relatively adaptable versus probably catastrophic rates of global change.

Estimating the rates of global warming in the next century is very controversial because of the uncertainties associated with these multiple interacting feedback mechanisms. Thus, applying the same climate models used to forecast the future to the mid-Cretaceous or to the ice age–Holocene transition provides some measure of validation. Scientists have done just that, and what they find is a relatively consistent gross picture between the performance of the models on paleoclimatic changes of past geologic eras and their projections of future climate changes. This is valuable circumstantial evidence, but it cannot confirm or deny the model's detailed regional projections.

What would happen if CO_2 were to double in the next fifty years? Scientists like Martin Hoffert at New York University and Curt Covey at Lawrence Livermore National Laboratory have looked at the climate versus CO_2 and methane differences between the height of the last ice age and the present era. They concluded that these differences were best explained if doubling CO_2 warmed up the Earth between 2° and 2.5°C (in the middle part of the current uncertainty range).[16]

The ice cores have shown (see figure 2.3) that climate, CO_2, and methane concentrations were relatively stable for roughly the past ten thousand years—the era of human civilization. That near-constancy in chemical composition of the greenhouse gases held up until the last two centuries, the industrial era. During the Holocene, recent ecosystems and habitats settled into the forms we know, following the 5°C globally averaged temperature rise and a 100-meter sea level rise that marked the five-thousand-year transition from last ice age to the current interglacial. It took nature roughly five thousand to ten thousand years to transform the landscape from ice over much of North America, Europe, and high-latitude seas to more current conditions, in which permanent ice is predominantly in polar seas and continents and high mountains. Since this transition coincides with about a 5°C global warming, we can estimate that natural rates of temperature change on a sustained, global basis are thus about 1°C per millennium. (Remember that number—we will return to it many times later on.)

Those changes, I have already said, were large enough to have radically shifted where and in what communities species lived. They may also have contributed to the extinction of such animals as mammoths and saber-toothed tigers.

GAIA OR COEVOLUTION?
....

So for some variables and scales, life has helped to stabilize climate change. Yet, in the transition from interglacial to ice age and back again, life seems to serve the function of accelerating rather than diminishing the climate change. These complexities led me in the 1980s to suggest an analogy to a biological process named by the ecologists Paul Ehrlich and Peter Raven a dozen years earlier.[17] Their research documented how the co-presence of two interacting species can lead to evolutionary paths that are both different because of their mutual interactions. They called this *coevolution*.

I saw an apt analogy, in that climate and life have coevolved. In other words, both life and the inorganic environment, including the meteorological elements, have followed evolutionary paths different from what otherwise would have happened over geological time had the other not been there. Coevolution does not require negative any more than positive feedbacks, just mutual interactions. And the Earth's fossil and sedimentary records certainly bear witness to such interactions.

Finally, if we humans are allowed to consider ourselves part of life, that is, the living natural system, then it could be argued that our collective impact on the Earth—what some analysts have referred to as "industrial metabolism" and the new science of industrial ecology—may very well also be an important coevolutionary factor in the future of the planet. (Whether that is good or bad is a value issue that will be discussed near the end of the book.)

The current trends of increase in human population, demand for higher standards of living, and use of technology and organizations to attain these growth-oriented goals all contribute to by-products that many economists call residuals—but that the rest of us call pollution.

While none of these natural planetary-scale experiments from geological history is precisely equivalent to the human-caused global change experiment now under way and thus cannot offer conclusive proof that our forecasts are right, they all add bits of circumstantial evidence hinting that current projections are at least reasonably plausible. They certainly reaffirm my assertion that in order to make the critical projections of future climatic change needed to understand the fate of ecosystems and us on this Earth, we must dig through land, sea, and ice to uncover as much of the geological, paleoclimatic, and paleoecological records as we can. Unfortunately, some shortsighted political interests consider it expedient to cut budgets for such seemingly esoteric work.

These records are the libraries of our natural history. They provide the backdrop against which to calibrate the still-crude instruments we must use to peer into a shadowy future, a future being increasingly influenced by one species: humans.

··
WHAT CAUSES CLIMATE CHANGE?

Modeling gives us a tool to make predictions about future changes or to help us explain important aspects of what happened in the past. The latter is crucial to test the models' performance against data describing actual paleoclimatic events. Such tests help scientists figure out how to use the information from these models and how to verify their forecasts. Then we will be better able to evaluate a number of public policy issues confronting us as we enter a new century.

Unfortunately, those models are not yet well calibrated for conditions very different from known climate patterns, which may not encompass all future conditions. Thus, we have to search constantly for ways to test models further. The best physical laboratory we have for such testing purposes is not a lab built of glass and steel but the very Earth itself, particularly our knowledge of its ancient periods.

FLUCTUATION OR TREND?
······································

Climate can be simulated on a range of different scales, from the tens of millions of years (for example, on the time scale of the Cretaceous) to a hundred-thousand-year scale, with

ice ages and interglacials alternating, or on a scale of every
few years.

Many scales are needed to understand and credibly forecast global changes. The global change question relevant to
Earth systems science is primarily about human-induced climate change. Global thermometer records, when suitably
averaged, show a warming of about 0.5°C (0.9°F) since the
middle of the nineteenth century (figure 3.1).[1] A few people
still argue that this warming trend, especially the record
warmth of the 1980s and 1990s, is just a natural fluctuation.
So perhaps it might be helpful to look at the kinds of characteristic changes that can be identified in various time series.

One kind of variation is periodic change in which the
time series oscillates up and down around some mean value.

THE RECORD OF PAST GLOBAL WARMING

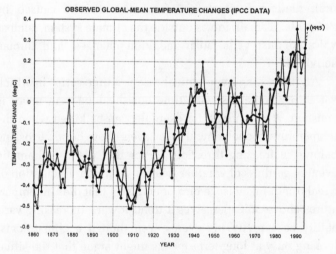

OBSERVED GLOBAL-MEAN TEMPERATURE CHANGES (IPCC DATA)

FIGURE 3.1

The combined annual land, surface air, and sea-surface temperatures (°C) from 1861 to 1995 relative to (that is, an anomaly from)
the 1951–1980 average for the globe, as constructed by the Intergovernmental Panel on Climate Change. The data show a century-
long warming trend of about 0.5° C (0.9° F). The years from 1981 to
1995 contain most of the warmest years on record.

Also possible is an impulsive change between two long-term means. When a volcano erupts, for example, throwing sulfuric acid aerosols in the stratosphere that block some sunlight and cause a rapid cooling below, that's an impulsive change. The surface cooling effect typically remains intense for a year or so, and then temperatures will "sawtooth" back up again in a warming trend over a few years. This is what happened in 1991 when Mount Pinatubo erupted in the Philippines, and the temperature effect can be seen in the figure.

There also can be shorter-term downward trends embedded in longer-term uptrends. The global surface temperature has been experiencing an overall upward trend in the last hundred years. Superimposed on this trend are accompanying "bounces" in temperature—both from year to year and over decades. There are vociferous arguments among researchers and climate watchers about whether these bounces are natural, random fluctuations or whether they're caused by definable, if small, forces outside the climate system, such as volcanic dust veils, solar radiation changes, and human activities.[2]

An interesting hypothetical case is a time series in which variability increases with time but the long-term mean is constant. A corn plant, for example, can be killed when the temperature cools below 0°C for even a few hours. Crossing below the freezing threshold is, for that plant, a dramatic event, regardless of whether it is just a random fluctuation or a real trend. Likewise, a bird or an insect that dies in temperatures above 30°C perceives a trend toward increasing variability as a pretty big event, even though a climatologist looking only at long-term means might argue that the situation reflects no climatic change. Even a few days of extreme heat can kill vulnerable people such as the elderly or the poor, as was sadly evident in the record heat waves that struck Chicago in July 1995.

Scientists are always looking for reasons behind a variation, and if those reasons are credible, it helps them distinguish change from fluctuation. The climate, as I have noted, was very varied in the past: There were ice ages, ice-free epochs lasting tens of millions of years, and even a billion or two years with little or no oxygen in the atmosphere. Compared to today, continents were in different places, the amount of energy from the sun was different, and the composition of the atmosphere was different. In other words, there have been natural "experiments" in the past that have seen extremely large-scale changes—even bigger changes, in many cases, than humans could ever effect by doing something to the chemical composition of the atmosphere in the next few decades. However, the natural rates of those changes usually—but not always—have been extremely slow relative to what people may force to happen.

To predict climate, we need to go beyond attempts at validating tools. We also need to identify and analyze what factors compel changes in climate, factors called "climate forcings."

CIRCULATION
·····················

The shape of the Earth's orbit, which controls the amount of sunlight that reaches the Earth at a given place and time, is a climate forcing. This solar heating forces the change of seasons, for example. The basic atmospheric circulations are driven by solar forcing. As sunlight comes in, some of it is immediately reflected back toward space—mostly by clouds, deserts, and ice caps. The reflectivity of the Earth is called the *albedo,* and it determines the amount of solar energy absorbed. Satellites measure albedo at about 30 percent for the Earth as a whole.

Because of the Earth's spherical shape, 50 percent of its surface area is between the 30° parallel North and the 30° parallel South. But much more than 50 percent of the sunlight is intercepted in these tropical and subtropical latitudes because of the geometry that places sunlight essentially overhead in the tropics but only at a glancing angle at high latitudes. As a result, the tropics are heated to excess, while the poles receive little heat.

But if the only thing controlling the climate were solar radiation, the extra heating at the equator would make it even warmer than it is, and the absence of sunshine at the poles in winter would make them even colder than they are. So there must be other processes at work. An obvious one is the heat that is constantly being transferred around the planet by fluids in motion, especially the atmosphere and the oceans.

Warm air rises in the tropics and moves upward and outward into colder areas and sinks thousands of kilometers away, toward the poles. As a current of heated air moves up and over the surface, there's also a return flow moving alongside or underneath, back toward the equator. This circulation is known as a Hadley cell. Add to that another complication, the fact that the world is a spinning ball, which deflects the flow of the moving air.

If you were riding along with a parcel of air moving first upward and then toward the poles, you would appear to deflect to the right in the Northern Hemisphere, and to the left in the Southern. You wouldn't actually be being deflected; it would only appear so relative to the spinning Earth below you. This is why cyclones rotate counterclockwise in the Northern Hemisphere and clockwise in the Southern. Cyclones have lower pressure at their centers than at their edges. Thus, air rushing in toward the lower-pressure center is deflected to the right (counterclockwise) in the Northern Hemisphere and the left in the Southern Hemisphere. The spiral-like appearance of such storms on satellite photographs arises from the combination of these deflections and the friction from winds rubbing on the Earth's surface. This is called a Coriolis effect, after Gaspar de Coriolis, who described the deflections in a mathematical equation centuries ago. Thus we have westerly (that is, blowing from the west) winds in mid-latitudes in both hemispheres because of the rising hot air in the tropics that deflects according to Coriolis forces.

Winds are produced by the temperature differences between one place and another in the atmosphere. That temperature difference, in turn, creates density and pressure differentials, which create rising air, wind currents, and so forth. The summertime jet stream is relatively weak; in the winter it's much stronger because the poles cool down but the tropics stay relatively warm all year long. Thus, temperature contrast between high and low latitudes is greatest in winter, the Hadley cell is stronger, more air and heat are transported poleward, and the circulation is more vigorous and the jet stream more variable and located closer to the equator.

When the large-scale circumpolar winds reach a certain velocity on a rotating body, they can become unstable. And if the jet streams are unstable, eventually they're going to break into high- and low-pressure eddies, also known as synoptic weather systems. The atmosphere obeys the physical laws of conservation of mass, momentum, and energy. These laws can be expressed as a set of equations whose solutions can simulate mathematically how the traveling weather systems work—the very paradigm Richardson tried to introduce in the 1920s. These simulations explain why the mid-latitudes typically experience weather patterns that change every few days, and why the tropics—and sometimes the mid-latitudes—have weather that is constant for months at a time.

The position of the jet stream is dramatically important for local and regional climatic conditions, since it steers storms and separates tropical from polar air masses.

You've no doubt heard about the "monsoons" of Africa, South America, and Asia (and a relatively weak one in North America). Oceans change their surface temperatures by only a few degrees from winter to summer because they're so thermally massive—they have what scientists call a large heat capacity, or thermal inertia. But land areas, with much less thermal inertia, can change temperatures seasonally by tens of degrees. Thus, the centers of the Asian, African, and South American continents really warm up during the summertime relative to the surrounding seas. The heated air over the land mass rises. As it does, something has to fill in: air blown in from over an ocean, laden with moisture. The result is summer monsoon: rainfall that sustains the natural and human ecologies of the regions.

Another common pattern of ocean temperatures is the cold, upwelling waters off the west coasts of continents. The reason for this is that the winds blow over the oceans, and friction creates currents. Along the west coast of North America, winds are typically coming from the northwest, seemingly pushing the ocean against the shore. But what is actually happening is a Coriolis force in the ocean deflecting the flow of water to the right in the Northern Hemisphere. That is, wind coming from the northwest in the Northern Hemisphere causes oceanic currents to deflect to the right, in effect pushing water away from the west coast. As the surface ocean water is deflected toward the southwest, that is, away from the coast, water from below—which is much colder—fills in. That's why you often need a wet suit to swim off the California beaches even in midsummer. This upwelling water is nutrient-rich and supports a diverse and productive marine ecosystem.

In addition to monsoon rains and cold upwelling currents off North and South America, climatologists and oceanographers also study the effects of a phenomenon called El Niño, which means, literally, "the child." This is a reference to the Christ child: El Niño is a recurrent phenomenon that is most noticeable around Christmastime. Every few years, atmospheric winds and an internal wave in the ocean, related to a massive sloshing back and forth across the equatorial Pacific, contribute to making the waters off

Peru uncharacteristically warm, while they cool at the western end of the tropical Pacific. The warming waters off the coast of Peru warm the atmosphere. The heated air rises, reversing the normal situation of sinking air over cool, upwelling waters. The winters of 1983 and 1995 were good examples. The warm sea-surface temperatures in the eastern Pacific altered wind patterns and steered storms southward into California, causing floods. Also, such wind changes have a feedback on sea-surface temperatures—part of a set of processes known as air-sea interactions.

Reversing the normally cool, upwelling waters in the eastern tropical Pacific leads to torrential rainfall in Peru, but also drought in Australia and even fires in New Guinea, which normally experiences a humid rain forest climate. What's more, El Niño has repercussions throughout the globe. The fluctuation between normal and El Niño circulation patterns is the so-called southern oscillation signal, which typically happens every five years or so. But from 1990 to 1995, El Niño–like conditions persisted—"the El Niño that wouldn't die," some called it. Was this a fluke or a climate change we'll have to live with[3]? Computer models of atmosphere, oceans, and coupled sea-air models are just beginning to simulate these factors successfully. This is a prerequisite to knowing whether global changes such as increased greenhouse gases might also affect the important El Niño phenomenon. As of today, strange phenomena like persistent El Niños are still unexplained, even if their consequences are all too well understood.

INTERNAL OR EXTERNAL CAUSES

In talking about reasons for climate change, I already pointed out the need to distinguish two basic categories: external and internal causes. "External" means arising outside the system and not influenced much by changes in the system—although external processes don't have to be physically external to the Earth (as is the sun). If our focus is on change in the atmosphere on a one-week time scale (the weather), then the oceans, land surfaces, biota, and human activities that produce carbon dioxide are all external: They are not influenced much by the atmospheric change on that short a scale. But if we are focusing on hundred-thousand-year ice age interglacial cycles, the oceans and ice sheets are part of the internal climatic system and vary as an integral part of the Earth's climatic systems. So which components are external and which are internal to our climatic system is not absolute, but depends on the time and space scales covered as well as the phenomena to be considered.

These are just a few examples that illustrate the complexity of deciding which factors to include in computer models as part of the internal climate system and which are external to that system. The importance of this internal/external-cause debate was identified in the 1960s by the MIT meteorological theorist Edward Lorenz,[4] a pioneer in the discovery of chaos theory.[5] Lorenz noted that a complex, so-called nonlinear system can have several types of behavior. *Nonlinear* means that the response of a system to some forcing is not a simple multiple of the strength of the forcing. For instance, if a nonlinear system that reacts with one unit of response to one unit of force were pushed with two units of force, it might respond with six (or one-half) units of response (or it might break—a very nonlinear response). Two aspirin tablets can cure your headache, but downing the whole bottle can kill you—the ultimate nonlinear response!

One mode of reaction is called "deterministic." This means that the system responds to forcing in a one-to-one way (even if nonlinear). That is, a given push causes a determinable response, and a double push causes another: There is direct cause and effect. For example, a volcanic dust veil that reflects 1 percent of the solar energy back to space will cause a unique cooling that is determinable in principle. A 2 percent reflection causes another (not necessarily linear, but still determinable) unique cooling response.

Another type of system behavior is "stochastic," meaning that the system will behave according to some statistical rules. For example, a pair of dice are not deterministic, since no relationship can reliably predict the outcome of any individual roll: Every time you roll them the odds are the same. Only a "statistical distribution" that gives the probability of any combination of faces for dice can be determined, at least in principle. Many weather systems exhibit such stochastic behavior, a factor that underlies the use of precipitation probabilities in weather forecasts.

Lorenz added a new kind of system behavior, later dubbed "chaos theory" by mathematicians. He suggested that some nonlinear systems are neither deterministic nor stochastic. There is a tendency for such systems to cluster around certain states that Lorenz called "strange attractors"—the ice ages and interglacials, in this metaphor. Many other examples of chaotic behavior in nature have been identified, ranging from the trajectories of toy balloons in space to irregular heat rhythms.

A debate rages in the scientific community over whether the climatic record results from external or internal causal factors, and whether this complex natural system is deterministic, stochastic, chaotic, or all of the above in different circumstances.

External causal factors and deterministic systems imply predictability. As an example, there is a sunlight detector at the Mauna Loa Observatory, located more than 3,000 meters above the Pacific on the big island of Hawaii. The detector typically determines that about 93.5 percent of the radiant energy from the sun that impinges on the top of the atmosphere actually reaches the surface at that very pristine location. In 1963 there was a noticeable drop—a couple of percent—of the sun's energy reaching Mauna Loa's detector.

That was the result of the eruption of Mount Agung on Bali, which threw sulfur dioxide into the stratosphere, where it photochemically converted to sulfuric acid particles, spread worldwide, and then slowly fell out over about a five-year period. This dust veil reflected away from the lower atmosphere an extra percent or so of the solar energy. The Earth should have gotten cooler, and indeed it did by a few tenths of a degree Celsius (see figure 3.1). Volcanic sunsets are spectacularly apparent because the high-altitude aerosol particles catch the just-set sun and the sky rebrightens, heavy on purples. In 1983 a volcanic mountain in Mexico called El Chichon erupted, throwing off a sizable chunk of mountaintop. The ash itself didn't cause significant climatic change, because it fell out of the lower atmosphere within weeks— wreaking havoc with the local residents. It was the sulfur dioxide sent to the stratosphere in the volcanic blast that was the real factor in climate change, also apparent in figure 3.1.

As a probable result of the explosion of Mount Pinatubo in the Philippines in 1991, globally averaged surface temperature in 1992 and 1993 was about .20°C colder than in the previous years. In fact, 1992 was the first year in half a dozen that had not registered in the record-high grouping. As the dust cloud cleared, 1994 and 1995 returned to record-high levels.[6]

Land use is another external climatic forcing that has to be accounted for, and there are many studies now being done using models to determine what could happen to the climate if humans were to deforest the Amazon at a rapid rate, for example. Forested area evaporates more water than cleared area, because trees have roots that tap into moisture deep in the soil. Leaves breathe in and out CO_2, O_2, and water vapor through tiny openings called stomates. The stomates in the leaves open up to take in CO_2 for photosynthesis, and they then let out moisture and O_2. Deforestation changes rates of evapotranspiration, which is an important component in feeding moisture to the atmosphere. CO_2 concentration in the atmosphere itself helps determine how long the stomates remain open, and this too influences evapotranspiration. If deforestation is accompanied by biomass burning on a large scale, then the smoke creates an aerosol that can alter temperatures, rainfall, and cloudiness. If human activities or natural processes modify the land surfaces and the biota, this can change the very nature of the weather upon which the biota depend. That's what is called "biogeophysical feedback," and it represents another set of internal processes that need to be accounted for in Earth systems modeling.

Runoff from land surface is linked to deforestation. Measurement in North Carolina showed more runoff after deforestation, because there is less vegetation to keep the soil intact and thereby to hold soil moisture. In addition, since there is less evapotranspiration from bare soils or grasslands than from forests, more water is left to run off. This can lead to downstream flooding, especially if soils erode enough to greatly increase runoff. However, the nonlinear climate system can be very complex. For example, if the deforestation takes place on a large enough scale, it may reduce evapotranspiration enough that it will decrease precipitation—and thus total runoff can go down, even if the deforestation increased the percentage of precipitation that becomes runoff. This scenario has been suggested for the Amazon. Flood control is one of the so-called ecosystem services provided for free by forests. Others include natural pest control, waste recycling, and the nutrient cycles.

The carbon cycle is in many ways linked to most of the factors, external and internal, I have discussed. It involves the greenhouse effect, photosynthesis, respiration, and decay—all natural processes. And, of course, the burning of fossil fuels and deforestation are global changes directly affecting the carbon cycle. The stocks and flows of carbon are mediated by the weather, and we know that humans are deeply in the game. The evidence is overwhelming that the 10 percent increase in carbon dioxide directly measured at Mauna Loa Observatory and other remote places from pole to pole since 1957 is due to human activities, as is the 25 percent increase found at ice cores in both poles since preindustrial times (see figure 2.3).

Which of all these internal and external factors is the most important influence on climate?

Clearly, changes to very slow cycles like carbon stored in soils isn't going to affect next year's weather, which is associated with more rapid changes like ocean-surface temperature patterns. A volcanic eruption is a very important external forcing factor for one or two years' average global temperature. At the time scale of a century (about how long it will take humans to double CO_2 or wipe out natural forests), volcanic dust veils appear just like short-term noise. So the answer to the question is an unsatisfying "it depends on which factor is most dominant relative to the scale of forcings and the characteristic response times of various subcomponents of the Earth system." Moreover, Earth system responses to various human disturbances could be any combination of several factors.

All of this explains why forecasting is still an inexact business. We have a system that is pushed and pulled—over differing periods of time—in myriad ways. We think we know fairly well which variables are internal to the system for a given time span and which are external, but we are not completely sure how each has influenced the system.[7] We know climate has nonlinear components, but we are not sure the extent to which each aspect is deterministic, stochastic, or chaotic. Volcanic dust veils and (probably) greenhouse gases drive largely deterministic responses. The seasons are largely deterministic and predictable responses to the Earth's orbital geometry, but the difference of a specific winter from the long-term average winter condition is, at best, only partly predictable.

Very often, critics deride climate models because the chaotic, unpredictable nature of the atmosphere renders weather forecasts unskillful in principle after a week or two. "If you can't forecast the weather accurately after two weeks, how can you dare pretend to forecast the climate—the long-term average of weather—two decades into the future?" is a typical tirade from some so-called climate contrarian critics.[8] However, just because we can't credibly forecast the sequences of faces on a pair of dice after many rolls doesn't prevent us from skillfully forecasting the probabilities of each combination of faces, nor does it preclude us from reliably forecasting an alteration to those probabilities if we had loaded dice and knew how they were rigged. Computer modeling is our only available tool to perform what-if experiments such as the human impact on the future (loading the "climatic dice" in this metaphor), and you can see it is tricky and easy to misrepresent since there are so many legitimate areas of uncertainty as well as many well-established facts.[9] But the very rapid rates of global change forcings leave us little choice but to use these evolving tools[10] based on known physical and biological principles to enhance our understanding, improve our forecasting skills, and inform the policy process about possible outcomes that will affect life on our planet in the next century.

CHAPTER 4

MODELING HUMAN-INDUCED
GLOBAL CLIMATE CHANGE

In 1628, the king of Sweden, Gustav Adolf of the House of Vasa, was eager to step up ship construction. He wanted a large fleet of warships with which to attack Europe.

At least one ship, named the *Vasa,* was built and launched in August of that year. Sixty-four bronze guns and a crew of 130 men were on board when the *Vasa* set out on her maiden voyage. Suddenly, before the ship left the harbor, a squall appeared and forced it to heel to port so far that water flooded in through the lower gunports. The ship sank in the harbor with its sails up and its flags flying. Fifty men died.

For more than three hundred years, the *Vasa* sat at the bottom of the Stockholm harbor in 100 feet of brackish Baltic water. She was raised in 1961 and found to be virtually intact, for the water's salinity was unfavorable for the destructive marine boring clams. One of the marine archaeologists who helped excavate the Swedish warship was Anders Franzén. He wrote in 1962 that there was no evidence to suggest that the *Vasa* had been badly designed or improperly sailed. "It is reasonable to assume," Franzén says, "that the cause of the catastrophe was an incorrect division of the guns, ballast, and other heavy weights on board."[1]

Perhaps the *Vasa* would not have capsized and sunk if the engineers had thought to build a scale model of the ship and to test its stability in winds with the heavy loads in different places. Such a model might have revealed that the position of the guns would create an unstable relationship between the ship's center of gravity and its center of buoyancy. Shipbuilding today depends not only on physical replicas of ships used as laboratory test models, but also on mathematical models in which the shape and weight of the vessel are manipulated in equations stored in a computer memory bank. These models simulate the performance of real ships on—and under—the high seas. Engineers and scientists build models—both mathematical and physical ones—primarily to perform tests that would be too dangerous, too expensive, or perhaps impossible to run with the real thing.

To simulate the climate, a modeler needs to decide which components of the climatic system to include and which variables to involve. As noted earlier, if we choose to simulate the long-term sequence of glacials and interglacials (the period between ice ages), our model needs to include explicitly the effects of all the important interacting components of the climate system operating over the past million years or so.

The problem for Earth systems scientists is separating out quantitatively cause-and-effect linkages from among the many possible internal and external factors. It's a controversial effort because there are so many subsystems and so many forcings operating at the same time. With such complexity, it's easy to pick fights if you don't like the results. But, as we'll see, despite the contention, models can be tested against reality with increasing confidence in their general conclusions.

So how do we do it? First, scientists look at observations of changes in temperatures, solar radiation, ozone levels, and so forth. This allows us to identify correlations among variables. Correlation is not necessarily cause and effect—just because one event follows another, it doesn't mean it was caused by it. For confident forecasting, we not only have to demonstrate the relationship but also explain how and why it happened. Especially for cases where unprecedented events are being considered, a first-principles, rather than a purely empirical, approach is desirable. However, observations yield correlations among variables that can lead to a hypothesis of cause and effect—"laws"—that can be tested against yet additional observations. The testing often involves comparing simulations with mathematical models on a computer against a variety of empirical observations—present and paleoclimatic.

That is how the scientific method is typically applied to climate models. When the simulations of a model, or set of linked models, appear plausible, they can be fed "unprecedented" changes such as projected human global change forcings—changes that may not have happened before—and asked to make projections of future climate, ozone levels, forests, species extinction rates, and so on. This is called "sensitivity analysis," since the model is used to estimate the sensitivity of the climate to a wide range of what-if events. The models become the testing laboratory for planetwide experiments we (I hope) would not want to perform with the real Earth as the laboratory.

The most comprehensive weather simulation models produce three-dimensional details of temperature, winds, humidity, cloudiness, and rainfall all over the globe. A weather map generated by such a computer model—known as a general circulation model, or GCM—often looks quite realistic, but it is never faithful in every detail. Large-scale patterns at hemisphere to subcontinental scales are usually more faithfully simulated than regional or local details.[2] To make a weather map generated by computer, we need to solve six partial differential equations that describe the fluid motions in the atmosphere and the laws of mass and energy conservation. These are known in meteorology as the "primitive equations." In principle, there would seem to be no problem, since we know from centuries of experimentation that those equations work; that is, we know that they describe fluid motions and energy and mass relationships. Why, then, aren't the models perfect simulations of the atmospheric behavior? Two answers stand out.

One is that the evolution of weather from some starting weather map (known as the "initial condition") is not deterministic beyond about ten days—even in principle: An event on one day cannot be said to determine precisely an event more than ten days in the future, all those commercial long-range forecasts such as *The Farmer's Almanac* notwithstanding. (Remember, anyone can forecast, but proving—not just claiming—a forecast accurate is what takes up most scientists' time). But the chaotic dynamics discovered by E. N. Lorenz that preclude, in principle, accurate weather forecasts much past a week or so do not rule out, in principle, accurate forecasts of long-term averages (*climate* rather than *weather*). The seasonal cycle is absolute proof of such deterministic predictability—as winter reliably follows summer, and the cause of this effect is known with certainty—and the climate models simulate the seasonal cycle quite well.

The other answer to the imperfection of general circulation model simulations, even for long-term averages, is that nobody knows how to solve those six complex mathematical equations exactly. It's not like an algebraic equation, for which one can get the exact solution by a series of direct operations. There isn't any known mathematical technique to solve such coupled, nonlinear partial differential equations precisely. Just as Richardson tried in the 1920s, we approximate the solutions by taking the equations, which are continuous, and breaking them down into discrete chunks, which we call grid boxes. A typical GCM grid box is about the size of the state of Colorado horizontally and, on average, at least hundreds of meters of atmosphere deep in the vertical direction.

I've already noted that clouds are very important, that they reflect sunlight away and trap infrared heat, as well. But because none of us has ever seen a single cloud the size of Colorado, we have a problem of scale: How can we treat processes that occur in nature at a smaller scale than we can resolve by our approximation technique of using large grid boxes? For example, we cannot calculate clouds explicitly because individual clouds are the size of a dot in this grid box. But we can put forward a few reasonable propositions on cloud physics: If it's a humid day, for example, it's more likely to be cloudy; if the air is rising, it's also more likely to be cloudy.

These climate models can predict the average humidity in the grid box, and whether the air is stable—likely to be rising or sinking—on average. So then we can write what is called a parametric representation, or "parameterization," to connect large-scale variables that are resolved by the grid box (such as humidity) to unresolved small-scale processes or phenomena (such as individual clouds). Then we get a prediction of grid box–averaged cloudiness through this parameterization. So the models are not ignoring cloudiness, but neither are they explicitly resolving individual clouds. Instead, modelers try to get the average effect of processes that can't be resolved explicitly at smaller scales than the smallest resolved scale (the grid box) in the GCM. Developing, testing, and evaluating the performance of many such parameterizations are the most important—and controversial—tasks of the modelers, whether they use climate, ecology, or economic models.[3]

This brings us back to one of the most profound controversies in Earth systems science; it is also one of the best examples of the usefulness, and fragility, of computer modeling.

THE GREENHOUSE EFFECT
....

If the Earth absorbed radiation from the sun without giving an equal amount of heat back to space by some means, the planet would continue to warm up until the oceans boiled. We know the oceans are not boiling, and surface thermometers plus satellites have shown that the Earth's temperature remains roughly constant from year to year (the 0.5°C warming trend in the twentieth century notwithstanding). This near constancy requires that about as much radiant energy leave the planet each year in some form as comes in. In other words, a near equilibrium, or radiative energy balance, has been established. The components of this energy balance are crucial to the climate.

All bodies with temperature give off radiant energy. The Earth gives off a total amount of radiant energy equivalent to that of a black body—a fictional structure invented by physicists that represents an ideal radiator—with a temperature of roughly 255K (–18°C or 0°F). The mean global surface air temperature is about 14°C (287°K, or 57°F), some 32°C (58°F) warmer than the Earth's black body temperature. The difference between the warm surface temperature and the Earth's radiative equivalent temperature (about a 32°C difference) is the celebrated greenhouse effect.

The term *greenhouse effect* arises from the classic analogy to a greenhouse, in which glass allows most of the solar radiation in and traps much of the heat inside. However, the mechanisms are different, for in a greenhouse the glass primarily prevents convection currents of air from taking heat away from the interior. Greenhouse glass is not primarily keeping the enclosure warm by blocking or reradiating infrared radiation, as the atmosphere does; rather, the glass structure largely is constraining the physical transport of heat by air motion. While some atmospheric scientists have thus advocated dropping this well-heeled term, *greenhouse effect* is both too entrenched and, even if inexact, not that bad an analogy to what the atmosphere does to trap heat near the Earth's surface. Ironically, perhaps, some environmental activists have also advocated dropping the phrase, but not because it is a physically inexact analog. Rather, they fear that since a greenhouse is a warm and friendly place for life, the term carries too benign an image with respect to the human enhancement of the atmosphere's heat-trapping capacity. They prefer *global heat trap*. As they say, you can't please everybody.

Although most of the Earth's surface and thick clouds are reasonably close approximations to a black body, the atmospheric gases are not. When the nearly black body radiation emitted by the Earth's surface travels upward into the atmosphere, it encounters air molecules and aerosol particles. Water vapor, carbon dioxide, methane, nitrogen oxide, ozone, and many other trace gases in the Earth's gaseous envelope tend to be highly selective—but often highly effective—absorbers of terrestrial infrared radiation.

Furthermore, most clouds absorb nearly all the infrared radiation that hits them, and then they reradiate energy almost like a black body at the temperature of the cloud surface—colder than the Earth's surface most of the time.

The atmosphere is more opaque to terrestrial infrared radiation than it is to incoming solar radiation, simply because the physical properties of atmospheric molecules and aerosol particles (including cloud droplets) tend on average to be more transparent to solar radiation wavelengths than to terrestrial radiation. These properties create the large surface heating that is characterized by the greenhouse effect, by means of which the atmosphere allows a considerable fraction of solar radiation to penetrate to the Earth's surface and then traps (more precisely, intercepts and reradiates at lower energies) much of the upward terrestrial infrared radiation from the surface and lower atmosphere. The downward reradiation further enhances surface warming and is the prime process causing the 32°C of natural greenhouse effect. This is not a speculative theory, but a well-understood and thoroughly tested phenomenon of nature.

The most important greenhouse gas is water vapor, since it is the most abundant trace gas and since it absorbs terrestrial radiation over many parts of the infrared spectrum. Carbon dioxide is another major trace greenhouse gas. Although it absorbs and reemits considerably less infrared radiation than water vapor, CO_2 is of intense interest because its concentration is increasing due to human activities. As we've mentioned, ozone, nitrogen oxides, sulfur oxides, some hydrocarbons, and even some artificial compounds like chlorofluorocarbons are also greenhouse gases. The extent to which they are important to climate depends upon their atmospheric concentrations and the rates of change of those concentrations.

The Earth's temperature, then, is determined primarily by the planetary radiation balance, through which the absorbed portion of the incoming solar radiation is nearly exactly balanced over a year's time by the outgoing terrestrial infrared radiation emitted by the climatic system to space. As both of these quantities are determined by the properties of the atmosphere and the Earth's surface, major climate theories that address changes in those properties have been constructed. Many of these remain plausible hypotheses of climatic change. Certainly the natural greenhouse effect is established beyond a reasonable scientific doubt, accounting for natural warming that has allowed the coevolution of climate and life to proceed to this point. The extent to which human augmentation of the natural greenhouse effect (that is, global warming) will prove serious is, of course, the current debate.

CAN MODELS BE VALIDATED?
....

This is a fundamental philosophical question. Strictly speaking, the logical answer is no, since, as already discussed, much of what humans are doing to force climate changes is unprecedented, and there is no precise empirical way to validate a model for conditions in which there is no exactly comparable test. But there is much that can be done in practice both to test model subcomponents and to evaluate overall model performance. While not perfect tests, they are more than adequate pieces of circumstantial evidence that allow subjective confidence judgments to be made about model performance.

There are many types of parameterizations of processes that occur at smaller scales than current models can resolve, and scientists debate which type is best. Are they an accurate representation of the large-scale consequences of processes that occur on smaller scales than we can explicitly treat? In forecasting climatic change, then, testing the validity of the model's parameterizations becomes important. In fact, we cannot easily know whether these parameterizations are "good enough." We have to test them in a laboratory. That's where the study of paleoclimates of the Earth has proved so valuable. We also can test parameterizations by undertaking specialized field or modeling studies aimed at understanding the high-resolution details of some parameterized process the large-scale model has told us is important.

Step back in time in America's heartland. Perhaps you've visited the sand hills of Nebraska? Although they're mostly grass-covered agricultural lands today, those hills were sandy between three thousand and eight thousand years ago because that part of the American plains was very dry then. The humid corn belt in Iowa and Illinois that we know today was much drier—what paleoclimatologists call a "prairie peninsula," a tongue of extra aridity a few hundred kilometers long.

Prior to the Holocene, about fifteen thousand to twenty thousand years ago, no corn belt would have been possible anywhere in the Midwest because it was too cold. Spruce trees typically found today many hundreds of kilometers north in Canada's boreal forests were dominant in the corn belt. As the ice gradually receded north and the climate warmed, the natural vegetation patterns underwent disorganization, movement, and transformation, settling down into the present patterns several thousand years ago, with grassland in the Western plains and hardwood forests in the Eastern plains and Northeast states.

Between about three thousand and eight thousand years ago, when summers were perhaps a few degrees Celsius warmer than today, the extensive aridity of the prairie peninsula was felt in the Mississippi valley. If a few degrees' warming were to recur—this time as a result of anthropogenic greenhouse gases—would the sand hills of Nebraska become sandy again?

Such a dramatic change would be stressful for agriculture as currently practiced in the central American plains or for the overall economy in that part of the hemisphere. Scientists would like to find out what caused the original warming and how the environment responded to it. If we knew this, could we then "hindcast" the aridity of the prairie peninsula, using the same tools we used to forecast the possibility of an enhanced greenhouse effect in the twenty-first century?

It is likely that changes in the Earth's orbit around the sun redistributed the amount of solar heating between winter and summer nine thousand to six thousand years ago relative to now—with some 5 percent more summer sunlight and 5 percent less in the winter. This could account for a few degrees' summer warming. I believe that warming from greenhouse-gas increases, which heat in both winter and summer, is probably not a good metaphor for what happened during the warm summers of the expanded prairie peninsula era. But does this mean that period has no lessons for the twenty-first century? Certainly not! If we can apply the same climate models we use to project future anthropogenic changes to the study of past natural changes, and if it appears to reproduce fairly well the patterns of past changes, then that evaluation process helps us to develop credibility in the model.[4] Once we've tested the model on large forced changes in the past, we can use it more comfortably to forecast forced climate changes in the future. We'll return to this six-thousand-year-old test case later on.

While geologists are digging up rock records, paleobotanists are in the field extracting cores of soil and lake sediment to bring back to the lab. There, scientists count the kinds of pollen grains left in each level of the core and typically determine their age by carbon-dating materials at that level.

Researchers figure out what tree or grass or herb pollen was there and in what relative abundances, date its age, and then infer from these relative abundances how the climate may have changed, according to such factors as different species preferring hot or wet climates, and so on.

The association between where the species were found and large-scale environmental factors such as temperature and rainfall is part of the subdiscipline called *biogeography*. Biogeographers can map at large scale (hundreds of kilometers) what types of vegetation groupings are likely to be present in a location simply by knowing its temperature and precipitation.

For example, tundra is predicted if summer temperatures are under 10°C (50°F), tropical rain forests if temperatures and rainfall are high, deserts if it is arid, and so on. Unfortunately, local factors such as soils, competition, and herbivory (plants being eaten by animals) limit such biogeographic "forecasts" to very general conclusions or rough approximations.

Researchers also look at marine or glacial deposits where

chemical composition of fossils or rocks or shells or ice can serve as proxy indicators of temperature and sea level. By taking samples from many locations, paleoclimatologists can look for signs of coherent patterns of change. Such patterns are needed for confident quantitative paleoclimatic reconstructions.

In all these ways, researchers have been able to deduce that there was an extended prairie peninsula in the Midwest, that it coincided with the middle part of the Holocene, and that there were many other changes taking place in the world at the same time. For example, fossil soils in the current deserts of Africa or India indicate that the monsoon precipitation zones in India and Africa were much wetter between five thousand and nine thousand years ago than they are now and than they were during the ice age. While there was relatively little change in the humid tropics six thousand years ago compared to now, the now-arid tropics underwent major changes. River runoff and the lake levels in north central Africa also were dramatically higher five thousand to nine thousand years ago.[5]

HOW ICE AGES (MAY) COME AND GO
....

If we look at more recent geological times, say, seven hundred thousand years ago to the present, a series of climatic cycles are evident. Every hundred thousand years or so, there is an interglacial that lasts ten thousand to twenty thousand years, followed by a transition into a very deep ice age tens of thousands of years later.

Most of the time between the interglacials and deepest glacials it has been colder than now. Interglacials tend to evolve toward maximum glacials more slowly: an eighty-thousand-year period of fluctuating ice buildup, then a ten-thousand-year, intense ice age maximum, and finally a very quick retreat of the glaciers (they retreated in about ten thousand years before a full interglacial period settled in). Paleoclimatologists call this a sawtooth pattern. There's much debate about what could have caused the slow buildup of the ice and the eventual more rapid decay. The following is a simple example of one possible sequence of events.

What was the extent of ice since the last ice age? Ten thousand to eleven thousand years ago, the northern half of the British Isles was covered by ice, and by eight thousand years ago there was very little of this ice left (see figure 2.2). In North America, ice extended from Long Island to Wisconsin and across much of Canada. Much of it didn't disappear until six thousand years ago.

How is it possible for enough ice to build up to produce an ice age? Many paleoclimatologists think—fortified by models—that the Milankovitch mechanism is a pacemaker of these cyles; that is, changes in the Earth's orbit alter the tilt of its axis, which modulates the amount of sunshine between winter and summer and equator and poles.[6] One conceptual account of glacial/interglacial cycles goes as follows: One winter there is an unusually heavy snow, one that doesn't melt completely in the summer because the orbital elements are favorable (meaning there is less summertime sunshine in Northern latitudes). The snow reflects away more of the sun's heat than trees, grasses, and soils would reflect, thereby dropping temperatures and making the next summer even cooler, a classic positive feedback system. Eventually snow builds up and compresses into ice, and the sheet of ice heads south with the colder weather. After fifty thousand years or so, ice has passed out of the Arctic to England, from Canada to Wisconsin. The enormous weight of the ice depresses the Earth's crust below it. Sea levels drop 100 meters as ocean water ends up on land as ice sheets.

How can the glacial period be reversed? More plausible conception: It gets so cold that the northern latitudes are deprived of heavy snows, halting the growth of the glaciers.

The weight of the ice sinks the bedrock below it, lowering the top of the ice sheet, exposing it to relatively warmer air. The Earth's orbit changes again, increasing summer sunshine, and the combination makes the glaciers recede. As more bare land is exposed and vegetation grows back, the Earth absorbs more heat and this positive feedback carries Earth rapidly (ergo the sawtooth pattern) into an interglacial period, after ten thousand to twenty thousand years of which the whole cycle starts anew.

When researchers build climatic models and put in these kinds of forcing and feedback factors, they can indeed reproduce that observed sawtooth pattern of ice age interglacial transitions in the computer output. However, successful paleoclimatic simulation with a model is still circumstantial evidence for global-warming validation exercises because we don't know for sure that the mix of these mechanisms just outlined operated in nature the way we've incorporated them into our model. For example, the one-hundred-thousand-year cycle that has been dominant for the past eight hundred thousand years is unlikely to be driven by changes to the eccentricity of the earth's orbit, because this one-hundred-thousand-year beat causes too small of a change in incoming solar energy. Recently, it was pointed out that a long-neglected variation in the tilt of the earth's orbital plane also occurs, and this variation exhibits a good match with the hundred-thousand-year cycle of ice ages for the past six hundred thousand years. But no clear mechanisms make this match look like much more than a coincidence right now. One interesting hypothesis suggests that the more intense ice ages—including the 100,000-year cycle of the past 600,000 years—were aided by the tectonic uplift of the Tibetan plateau. Obviously, ice age theory is not yet a closed issue. But there are enough consistencies among many aspects of paleoclimatic reconstructions and model simulations to lend considerable credibility to several basic ideas (see page 158, note 13).

THE CLIMATE OPTIMUM
....

Astronomers, I've noted, have shown that the Earth's orbit goes through twenty-thousand-, forty-thousand-, and hundred-thousand-year cycles in its orbital relation to the sun. Today, we're closest to the sun in January, but nine thousand years ago, we were closer in July. Ten thousand years from now, we'll reverse again. We know that the Earth's orbital changes don't alter the *total* annual amount of sunlight received by the Earth by more than a couple of tenths of a percent, but the orbital effect could change the latitudinal and seasonal distribution by as much as 10 percent, the so-called solar-orbital forcing. We are pretty sure that about 8 percent more sunlight came into the Northern Hemisphere nine thousand years ago in the summer.

With recent advances in computer modeling, modelers can now begin to simulate and explain these changes in past climates. They can take what we know about ice-cap changes, particles in the atmosphere, carbon dioxide, sea surface temperatures, and energy from the sun, and feed all these "forcings" into models. They then produce simulations of what the climate was like thousands of years ago.

Scientists studying this phenomena also studied fossil pollen in lake sediments and observed how spruce forests moved north. Then, using a climate model's prediction of paleoclimate change, plus another model for how forests change as temperature or rainfall changes, they connected climate science with ecosystem science. This allows us to predict what happens to ecosystems driven by model-calculated temperature and rainfall changes over time and to test the climate model by checking its predicted ecosystem changes against what was found in the field with fossil pollen. An international consortium of scientists from many Earth science disciplines collaborated over half a dozen years in a program known as the Cooperative Holocene Mapping Project (COHMAP).[7] Overall, their many climate and ecology models versus data comparisons are encouragingly similar with regard to broad patterns, but individual details are not always in very good agreement. At any one given location at any one point in time, these models aren't yet very credible in predicting specific details. Moreover, their forest models didn't include the direct physiological effects on tree growth of lowered CO_2 concentrations in the atmosphere at ice age times.

The fundamental question, then, becomes: Does the broad correlation that can be observed between model-predicted changes and natural changes over 160,000 years validate a quantitative cause and effect between temperature rise and greenhouse gases buildup in the last—or the next—hundred years? Not quite yet, because there still are other potential explanations. But agreements are consistent enough to argue that such cause-and-effect associations are quite plausible—I consider it 80 to 90 percent likely that there is a causal relationship between the twentieth-century warming trend and greenhouse-gas forcings. (See chapter 6 for other scientists' subjective probability opinions on this point.) The evidence is strong, but still circumstantial and thus not conclusive—a condition ripe for contention among special interest groups.

HAS A HUMAN-INDUCED CLIMATE CHANGE BEEN DETECTED YET?
··························

Thermometer records (see figure 3.1), I've noted, suggest about an 0.5°C warming trend over the twentieth century. Coincident with this is the augmentation to the concentration of greenhouse gases such as CO_2, CH_4, and N_2O. Many policy analysts and decision makers have asked whether this correlation is a coincidence or cause and effect. In short, has the human-induced climatic signal been detected in the observed temperature record? The answer to this question might seem straightforward, but is exceedingly difficult—a situation that spawns much controversy.

First of all, to "detect" some signal implies picking it out of a noisy background. The global average temperature record exhibits fluctuations of about 0.2°C from year to year and from decade to decade. Are they simply random noise, or forced responses to phenomena like aerosols from volcanic eruptions? I think the answer is both. The global temperature dips of a few tenths of a degree (see figure 3.1) for a couple of years after volcanic blasts in 1883 (Krakatoa), 1963 (Agung), 1983 (El Chichon), and 1991 (Pinatubo) are very likely to include a forced response to the stratospheric dust clouds spawned by those volcanoes, whereas most of the other year-to-year fluctuations in temperature are probably just "noise"—random, or "stochastic," internal oscillations caused by the exchange of energy and materials among the climatic subsystems: atmosphere, oceans, ice fields, soil, and biota.

What about the century-long 0.5°C warming trend? Could it be climatic noise? This is similar to asking whether one roll of a pair of dice that produces "snake eyes" (two one-dot faces—a 1 in 36 chance) suggests that the dice are loaded. Most of us would like to roll for a while to check out the odds. But in the case of the Earth and its climate, we don't have global coverage of thermometers for more than a century, so there are no direct measurements to tell us the odds of the "climate dice"—in this case, the probability that an 0.5°C century-long warming trend is a chance occurrence. Put more mathematically, we need to know the long-period natural variability (climatic noise) in order to see if the 0.5°C trend is comparable to or larger than this natural noise. If larger, we can be more confident that the twentieth-century warming is not a random event (we call this process "climatic signal detection"). However, even if we do establish a high probability of detection of climatic change, there is still more work to do before ascribing the detected change to human activities (we call this issue "climatic signal attribution").

Since we don't have direct observations of global surface temperature trends over, say, two thousand years (a span that contains twenty hundred-year periods), some have suggested that there is no direct evidence of climate change so far. Although strictly true, this is a highly misleading statement because there is a great deal of *indirect* evidence. For example, the width of tree rings is a proxy for climatic change, and scientists have archived thousands of tree-ring sequences going back thousands of years all over the globe. Other proxy "thermometers" include land-form changes from glacial movements, pollen abundance changes in lake sediments, and the chemical composition of old snow layers in glaciers. While not being precise global temperature indicators, taken together the proxy records suggest that an 0.5°C global warming (or cooling, for that matter) is a fairly unnatural event, occurring perhaps no more than an average of once in a millennium during our recent interglacial history. This indirect evidence strongly supports the notion that a real climate change has been detected in the twentieth-century thermometer records. That is why I believe it is 80 to 90 percent likely not to be a natural fluctuation. But how to attribute the cause of this trend?

To suggest that human increases in greenhouse gases caused the warming trend requires elimination of other potential causes (such as changes in solar radiant heat or volcanic eruptions). Again, direct evidence of solar energy output is available only for a relatively short period: the last two decades in which space-borne instruments flying above the distorting effects of the atmosphere have taken measurements. These measurements show that over an eleven-year sunspot cycle, there is a little change (less than 0.5 percent) in solar radiation, too little to explain most of the observed global temperature record.[8] Of course, there might have been larger changes in the solar output before we had reliable, space-borne measurements. This possibility has led to a noisy debate in which greenhouse skeptics argue that solar variability could have caused the observed warming trend (even though they have no direct evidence). Although I and most of my colleagues do not think it likely that the sun alone can explain the climatic changes of the past hundred years, neither can we rule this out at a very high probability, like 99 percent. It will take about ten to twenty more years of observed solar radiation and global surface warming (1995 was another record warm year), I believe, to assert with very high probability that we can attribute X amount of the observed century-long warming to human activities. That is, it will take decades for more certain attribution *if* we use only global averaged surface temperatures as our measure of change.

A police officer might find that several suspects have no alibi, but that hardly proves them guilty. Other direct evidence is sought, especially fingerprints. Are there "climatic fingerprints," too? The answer is a qualified yes. For example, carbon dioxide heat trapping causes climatic models to increase their globally averaged temperatures, but, in addition, stratospheric temperatures decrease, Northern Hemisphere temperatures warm up more than Southern, and polar latitudes warm up more than the tropics because melting ice and snow cause more sunlight to be absorbed, amplifying the high-latitude climatic signal. Thus, a pattern of change, or climatic fingerprint, is found in models in which CO_2 has been doubled. Then climatologists search observational records to see if this model-produced fingerprint occurs in nature. The results have been mixed. Observed warming has occurred, of course, but the Southern Hemisphere did not warm up less than the Northern, nor did extra polar warming match the model's projections. The stratosphere has indeed cooled, but by more than models forecast from greenhouse gas increases.

Greenhouse skeptics and their political allies loudly proclaimed the models were incompetent because there was no clear fingerprint in the observed record. But as a number of climate scientists (myself included) pointed out in response, it is only fair to compare the climatic change patterns in nature with those of climate models *if* the models are driven with the same set of external forcings that nature has been subjected to.[9] In other words, the models need to be driven not only with increased greenhouse gases, but also the potentially important regional cooling effects of aerosols generated by burning high-sulfur coal and oil (among other forcings, especially ozone changes and biomass burning, thought to be of lesser absolute importance) need to be included. Recent models driven by both global-scale greenhouse gas increases and regional patterns of sulfate aerosols show quite different fingerprints (climate change patterns) than models driven by greenhouse gases alone or by changes in the energy output of the sun by itself.[10] The bulk of these industrially produced aerosols reside in the Northern Hemisphere. Thus, since they reflect away a little bit of the sun's energy (especially in the summer), thus cooling the climate, one might expect the aerosols to offset somewhat the magnitude of warming from greenhouse gases alone, with the bulk of the offset occurring in the Northern Hemisphere. Indeed, recent model runs at the U.K. Hadley Center and the Lawrence Livermore National Laboratory in California with both CO_2 and aerosol forcings show slightly higher warming in the Southern Hemisphere, and little high-latitude amplification. The stratosphere still cools—closer to observations if the effects of ozone depletion are simultaneously included.[11] This climatic fingerprint pattern is tantalizingly closer to the observed regional and seasonal patterns of climate change from 1960 to 1990. This encouraging agreement has led the hundreds of scientists associated with the Intergovernmental Panel on Climate Change in 1995 to conclude, cautiously, that it is now likely that a real climate change has been detected and that at least some of it can be attributed to human activities. While acknowledging many remaining uncertainties, the Executive Summary of the IPCC Report concluded (after debating for days over the wording of this one sentence): "nevertheless, the balance of evidence suggests that there is a discernible human influence on climate."[12] Hours of debate alone attended the choice of *discernible*—a word not very distant from the one I used in my 1976 book, *The Genesis Strategy*: "demonstrable."[13]

Whether this new fingerprint evidence qualifies as a smoking gun or is simply an improbable coincidence will spawn debate for years. In the meantime, Laboratory Earth continues to grind out the answer—experimentally.

One final issue needs to be addressed in the context of climate detection and attribution. Until recently, climate modeling groups did not have access to sufficient computing power to routinely calculate time-evolving runs of climatic change, given several alternative future histories of greenhouse gases and aerosol concentrations. That is, they did not perform so-called transient climate change scenarios. (Of course, the real Earth is undergoing a transient experiment.) Rather, the models used to estimate how the Earth's climate would eventually look (that is, in equilibrium) after CO_2 was artificially doubled and held fixed indefinitely rather than increased incrementally over time, as it has in reality or in more realistic transient model scenarios.

Transient model simulations exhibit less immediate warming than equilibrium simulations because of the high heat-holding capacity of the thermally massive oceans. However, that unrealized warming eventually expresses itself decades later. This thermal delay, which can lull us into underestimating the long-term amount of climate change, is now being accounted for by coupling models of the atmosphere to models of the oceans, ice, soils, and biosphere (so-called Earth system models—ESMs). Early generations of such transient calculations with ESMs give much better agreement with observed climate changes on Earth. When the transient models at the Hadley Center in the United Kingdom and the Max Planck Institute in Hamburg, Germany, were also driven by both greenhouse gases and sulfate aerosols, these time-evolving simulations yielded much more realistic fingerprints of human effects on climate.[14] More such computer simulations are needed to provide overwhelming confidence in the models, but scientists are now beginning to express growing confidence that current projections are not the utter fantasy the critics repeatedly proclaim.

However, such a very complicated coupled system like an ESM is likely to have unanticipated results when forced to change very rapidly by external disturbances like CO_2 and aerosols. Indeed, some of the transient models run out for hundreds of years exhibit dramatic change to the basic climate state (such as a radical change in global ocean currents).[15] In 1982, Starley Thompson and I used very simplified transient models to investigate whether the time-evolving patterns of climate change might depend on the rate at which CO_2 concentrations increased.[16] For slowly increasing CO_2 buildup scenarios, the model predicted the standard model outcome: The temperature warmed more at the poles than at the tropics.

Any changes in equator-to-pole temperature difference help to create altered regional climates, since temperature differences influence large-scale atmospheric wind patterns. However, for very rapid increases in CO_2 concentrations, we found a reversal of the equator-to-pole difference in the Southern Hemisphere. If sustained over decades, this would imply unexpected climatic conditions during the century or so the climate adjusts toward its new equilibrium state. In other words, the faster and harder we push on nature, the greater the chances for surprises—some of which are likely to be nasty.

Fifteen years later, the IPCC concluded its Executive Summary with the following paragraph:

Future unexpected, large and rapid climate system changes (as have occurred in the past) are, by their nature, difficult to predict. This implies that future climate changes may also involve "surprises." In particular these arise from the non-linear nature of the climate system. When rapidly forced, non-linear systems are especially subject to unexpected behaviour. Progress can be made by investigating non-linear processes and sub-components of the climatic system. Examples of such non-linear behaviour include rapid circulation changes in the North Atlantic and feedbacks associated with terrestrial ecosystem changes.[17]

Of course, the system would be less "rapidly forced" if we chose as a matter of policy to slow down the rate at which human activities modify the atmosphere. I will take up this controversial issue at the end of the book.

CLIMATE SURPRISES
....

About twelve thousand ago, after warm weather fauna had returned to northern Europe and the North Atlantic following the long glacial period, there was a dramatic return to ice age–like conditions in less than a hundred years. This miniglacial is known as the Younger Dryas, after the widespread reappearance of the Dryas, a tundra flower. It lasted for some five hundred years before the warm, stable Holocene finally set in. What happened?

We don't know for sure, of course, but there are good hypotheses. Moreover, the so-called Younger Dryas climate signal was largely a regional change: the whole North Atlantic, including northeastern Canada and most of Europe. But the period saw dramatic ecological retrenchment in decades to ice age–like plants and animals over that region.

Globally, there is evidence of simultaneous changes, but these appear much less dramatic and no significant climate change is evident in Antarctic ice cores at this time in geologic history. Studies of fossil plankton remains in the sediments of the North Atlantic suggest that the warm Gulf Stream current was directed many degrees of latitude south, and that the overall structure of deep ocean circulation (sometimes called the oceanic conveyor belt) reverted to near ice-age form in only decades—a dramatic climatic change measurable in the lifetime of a single human.

The most plausible hypothesis for the Younger Dryas is a quick pulse of fresh water into the North Atlantic. Since fresh water freezes much more readily than salty water, it would permit the rapid establishment of sea ice cover, which could explain the dramatic chilling down of Europe about twelve thousand years ago. But where would such a rush of fresh water come from? The most likely explanation is advocated by the Columbia University geochemist Wallace Broecker:[18] Meltwater from the rapidly warming North American ice sheet accumulated in a giant great lake (called Lake Agassiz by geologists) whose Eastern bank was the remnant ice sheet; then the ice dam broke and a dramatic "meltwater spike" rushed down the Saint Lawrence Valley to the North Atlantic.

Recently, very controversial findings suggest that, at Greenland at least, dramatic flip-flops in both temperature (5°C in decades) and CO_2 occurred several times in the previous interglacial, some 130,000 years ago.[19] This period, up to now, was generally believed to be slightly (2°C) warmer on average and comparably stable to our interglacial period, the Holocene. The most popular hypothesis to explain the alleged flip-flops 130,000 years ago is a flip-flop of the North Atlantic conveyor belt circulation. These still-debated surprise climatic variations have led to an obvious and critical question: Could such a rapid change to the conveyor belt ocean current be induced today by pushing the present climate system with human disturbances like greenhouse gases or sulfur oxides? Could an anthropogenic global warming of 2°C—expected as a good bet in the decades ahead—trigger such surprise climatic instabilities as may have occurred in the North Atlantic region during the 2°C warmer interglacial 130,000 years ago?

The best way we can assess the risks from rapid human forcings on the climate is by comparing the models and paleo-data to try to figure out what happened in the past, from which we can estimate what the chances are of something significant happening again. Again, the form and likelihood of future surprises are speculations based on plausible, as yet unconfirmed, analyses. But the prospect of climatic surprises in general is chilling enough to lend considerable urgency to the need to speed up our understanding, slow down the rates at which we are forcing nature to change, or, better (in my value system), do both.[20] How much that might cost or who should pay will be addressed briefly later on. First let's consider how climatic changes might work with or against other human disturbances, and how such global changes might impact on natural ecosystems.

CHAPTER 5

BIODIVERSITY AND
THE BATTLE OF THE BIRDS

The ice age forests of central Europe and the Middle Atlantic states were heavily populated with spruce trees rather than the hardwood species of today, like oaks and maples. For a long time, it was believed that communities of species simply marched back and forth, chasing the ice caps north in melting periods and racing back south ahead of the ice as the planetary chill set in. That view had been popular since the time of Charles Darwin, who believed that whole communities of species simply migrated as a block with changing climate. In *The Origin of Species* he wrote:

> As the arctic forms moved first southward and afterward backward to the north, in unison with the changing climate, they will not have been exposed during their long migrations to any great diversity of temperature; and as they all migrated in a body together, their mutual relations will not have been much disturbed. Hence, in accordance with the principles inculcated in this volume, these forms will not have been liable to much modification.[1]

Such an idea is reassuring for the preservation of biodiversity if the climate were to change: The species would move together and survive. If whole biological communities simply migrated with climate change, then as long as free access to migration routes were maintained, species could adapt to climate change via migration with little loss of biodiversity (also known as extinction) as a consequence.

But that expectation of community migration has been sharply revised in the light of recent data dug from the soils and sediments of the Earth. The ecologist Margaret Davis, at the University of Minnesota, first suggested that individual species responded differently as the climate warmed.[2] Then, the consortium of scientists who call themselves the COHMAP (Cooperative Holocene Mapping Project) examined the fossil remains of pollen grains from many types of plants, going back to the height of the last ice age. They found that species like spruce or oak did indeed chase the ice cap northward as it melted, but not the way Darwin or most ecologists envisioned, as a march of intact communities. Instead, the COHMAP scientists discovered that during the transition from ice age to interglacial times, species moved at different rates and even in different directions from each other. One might say that the trees moved but the old forests disappeared, as the composition of trees, herbs, and grasses went through many unique or unfamiliar combinations during the transition—what is called a "no-analog" habitat, since nothing like it exists today.[3] These ecological rearrangements may have even contributed to the most recent extinction event: the disappearance of mammoths, the saber-toothed tiger, and other so-called charismatic megafauna at the end of the last ice age.

Such no-analog habitats also were unearthed by Russell Graham at the Illinois State Museum for small mammals responding to the retreat of ice some five thousand to fifteen thousand years ago.[4] His finding also contradicted Darwin. These should be disturbing findings for those who hold high confidence in the proposition that biodiversity could be preserved by migration in the face of climate change, since they show that normal associations of plants and animals can be disrupted during climatically induced transitions—even for relatively slow natural climate changes (recall from chapter 2 and figure 2.3 that the average sustained rates of global temperature change from ice age to interglacial times was only about 1°C per thousand years).

In other words, the "balance of nature," including predator-prey relationships and other competitive mechanisms that create seemingly stable distributions of species in an ecosystem,[5] could be seriously disrupted during a climate change, since it could cause different species to respond at different rates, thereby altering the biological community structure. This has led some ecologists to worry that the disassembly of biological communities caused by rapid (at sustained globally averaged rates of degrees per century as currently projected) climate change could exacerbate species extinction rates. Current rates of extinction are already believed to be thrust upward by habitat fragmentation, chemical pollution, and the introduction of non-native (so-called exotic) species into new habitats.

The University of Michigan ecologist Terry Root has studied the distribution of wintering birds in the United States, looking for any correlations with large-scale environmental variables like temperature or vegetation patterns. She found a remarkable number of such correlations like that on figure 5.1 for the eastern phoebe, a species of birds whose winter northern range limit has a close association with mean minimum January temperature; but this winged animal could migrate very rapidly in response to climatic changes.[6] Doesn't this suggest that global change may be irrelevant to such mobile species?

Root discovered that many species of wintering North American birds associate with both temperature and vegetation. Temperature often controls how far north birds could survive given their physiological tolerances, whereas vegetation of a certain type might be required for food, shelter, or

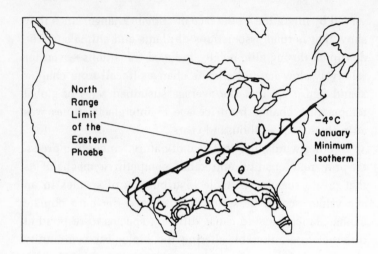

FIGURE 5.1

The average northern range boundary of the eastern phoebe in winter coincides closely with the line that represents the average January minimum (nighttime) temperature. Terry Root has found that many other birds wintering in North America also have their ranges constrained by climate factors. She suggests that climatic changes typically projected into the 21st century could seriously alter current bird communities.

nesting. Root further noted that birds that are physiologically constrained by low temperatures alone could migrate north when it warms, but those that are also restricted by habitat (for example, vegetation patterns) may have to wait centuries for their required vegetation to adjust to altered climate and photosynthetic reactions to the extra CO_2 in the air[7] and to migrate—if it can. In the interim, what is likely to occur is a tearing apart of the structure of ecological communities, alteration of predator-prey interactions, and the potential for ecological disorder during the few centuries it will take for climate to warm from as little as 1°C to perhaps as much as 10°C, and for the various species to respond individualistically to both climate and direct CO_2 changes. Such disruption to natural balances would likely enhance extinction, especially for the many species that have limited habitat ranges and are strongly associated with climate variables.

It already is a formidable scientific challenge to try to explain the range limits and abundances of most species today, even though they have had thousands of years of stable climate and CO_2 concentrations to adapt to. The fear that disruption of ecosystems could lead to the tearing apart of communities of plant and animal species, or even a scrambling of ecosystems as currently constituted, might be justified should global warming materialize at typically projected or higher rates.

DOES IT MATTER WHO'S IN A COMMUNITY?

I recently attended my thirtieth high school reunion. I was a bit anxious at the prospect of seeing hundreds of former classmates who had had three decades to gray, wrinkle, spread, and (with luck) mellow and mature. That some were unrecognizably different and other remarkably unchanged didn't actually surprise me much. What did was the realization of how very few people I could remember, even with a copy of my yearbook handy. There were a half-dozen former friends and twice that many casual acquaintances I remembered, but the hundreds of others rang virtually no memory bell. An unanswerable question popped into my head: Would I be very (or at all) different today if most of these classmates had been a year ahead of me, or a year behind? Were we a community of interacting people with a group identity that indelibly changed each of us or, rather, a random assemblage of individuals whose functioning (in this case, high school education) would have been largely unaltered if we had been brought together with an entirely different set of students? Do the particular species in a biological community matter to the functioning of the ecosystem, or only enough of certain types of families? If one tree species went extinct and another took its place, would the community be fundamentally different?

The concept of a species role in a biological community leads to another idea, that of ecosystem services. Both of these are somewhat controversial. An ecosystem is defined as a community of different species all interacting with one another and the physical and chemical environment.

An ecosystem's size can range from a microbial community in a cloudy drop of water to the world, depending on the particular system and time frame we wish to consider. The functioning of an ecosystem results in many interactions that, over time, lead to a condition that used to be called a "balance of nature." But few ecologists hold to that static view today. Rather, the phrase "flux of nature" is more often used to describe a dynamic equilibrium in which many individual species can fluctuate in numbers, and over long enough times go extinct, only to be replaced by newly evolved life forms. Yet somehow ecosystem services are maintained. Such services include primary productivity (photosynthesis, oxygen production, the removal of CO_2 from the air and its fixation into plant materials—the base of the food chain), recycling of wastes (from a vast array of decomposers), flood control (vegetated slopes dramatically reduce runoff relative to denuded hillsides), maintaining genetic resources (for food, disease resistance, and medicines), and water purification, to name a few. It is controversial whether human activities, which threaten to reduce biodiversity, would also threaten these ecosystem functions and, if so, whether humans could replace such free services of nature with sustained and affordable technological substitutes.

I also said the concept of biological communities is controversial. No one denies that populations of various species occupying (and interacting, in most cases) some specific place define a biological community. Rather, the debate is over whether that particular set of species has a unique attribute or function that would not be replicated by substituting different populations or species.

Two extremes are possible: (1) communities of species have no unique ordering, but are simply random associations; and (2) species interactions are so tightly bound that the community as a whole is greater than the parts that would be collectively disrupted by a disturbance to any one species. Examples of both extremes can be identified in nature, although most ecologists believe something midway between these extremes is more representative. Not all species are equally important to the functioning of a community. Some are able to expand or contract their population size with little effect on other members of the community, whereas others, called keystone species, play a pivotal role in community structure.

One well-known example of a keystone crisis was the near extinction of the U.S. West Coast sea otter by fur hunters. After their decline, a major disturbance propagated through the offshore marine community. Sea urchins, normally a principal food for otters, multiplied rapidly and in turn decimated the kelp forests leading to biologically impoverished, desertlike stretches of sea floor known as sea-urchin barrens. Only after controversial political pressures to restore the otter were successful did the urchin populations decline, the kelp grow back, and a new community of fish, squid, and lesser organisms reestablish themselves.

Many other such stories can be told, like the eradication of wolves in the western U.S. to protect livestock, leading to a rise in coyotes, followed by coyote-control programs that allowed a fox population explosion that threatened waterfowl populations. This has led to controversial proposals to reintroduce wolves.

It often is not clear in advance of performing the experiment in nature whether any particular community is loosely knit or tightly bound, let alone what the keystone species are in that community or whether there are critical population thresholds below which extinction is likely and reverberations will plague the community for a long time. Conservation biologists have drawn up rough rules of minimum population sizes and habitat areas below which extinction (locally at least, and globally if there is only one area left containing that species) is looming. Such formulas have also been used to forecast how human activities, most notably deforestation, will reduce future biodiversity. Since such forecasts have led to demands to slow the destruction of some habitats—a land use that may be economically advantageous to special groups—the formulas themselves have been under attack. Let us examine in a little more detail the basis for projecting species loss as a result of global changes like deforestation.

ISLAND BIOGEOGRAPHY: A BIODIVERSITY CRYSTAL BALL?

Large regions are often characterized by life zones such as tundra, boreal forests, deserts, grasslands, or tropical moist forests. These biomes, or life zones, often repeat themselves around the globe where climate conditions are appropriate—for example, tundra in cold, high altitudes or latitudes and a broad-leafed vegetation rain forest where it is hot and humid. The correlation between January nighttime temperature and the northern-range edge of many North American wintering birds found by Terry Root results from the physiological constraints birds experience where nights get longer and colder. They need to crank up their metabolic rate enough to quite literally shiver through the night, using up fat as fuel. This constitutes losing something like 10 percent of their body mass in one night! It must be replenished the next day or such birds will have to live further south or die.

Similarly, plants that populate various biomes also possess physiological characteristics that allow them to occur in certain geographic and climatic regions. The relationships between the distribution of species or life zones and large-scale factors like climate are known as biogeographic associations.

Another factor influencing the biogeography of species distributions is the size of a patch of habitat or the distance between such patches. Biogeographic rules relating the diversity (number of species) of a place to its climate, size, and isolation from other places have been sought by studying islands. Islands newly formed from volcanic action begin devoid of life, but soon become populated as plants and animals arrive with the winds (or, later on, hitch a ride on the bottom of ships, as the highly destructive zebra mussel did as it invaded North American waters from Europe in the 1980s).

The closer the island is to a source of new migrating species (like the mainland), the easier it is to survive the trip, thus the greater the number of species on the island. This is known as the "distance effect." The larger the size of the island, the greater the number of places or ecological niches different species can find to survive. In addition, more space allows more individuals, so more species will have population sizes above the minimum thresholds for long-term survival. Taken together, these mechanisms comprise the "area effect." In other words, the larger the island, the larger the number of species is likely to be. Eventually, if all external factors (which ecologists call "disturbances"), like climate change, fires, or bulldozers, are held constant, a dynamic equilibrium is established in which the rate of extinction of species already on the island is balanced by the arrival of new ones from the outside. The equilibrium part is that under these circumstances, the diversity remains about the same, whereas the dynamic aspect is that the roster of species is changing over time.

In 1963 Robert MacArthur and Edward O. Wilson used these ideas and lots of data to formulate a theory that today is a prime tool for forecasting how global change activities like deforestation could reduce biodiversity. This relationship, known as the theory of island biogeography, has profound implications for the impacts of humankind on four billion years of the Earth's natural history in which life and climate have coevolved. The theory has led to angry confrontations among ecologists using the theory to argue for humans to slow their assault on nature and development interests, who either value nonhuman nature relatively little or attack the theory as an exaggeration. Harvard University's Ed Wilson, in his award-winning book, *The Diversity of Life,* explained how their theory came about:

> We had noticed that faunas and floras of islands around the world show a consistent relation between the areas of the islands and the number of species living on them. The larger the area, the more the species. Cuba has many more kinds of birds, reptiles, plants, and other organisms than does Jamaica, which has a larger fauna and flora in turn than Antigua. The relation showed up almost everywhere, from the British Isles to the West Indies, Galápagos, Hawaii, and archipelagoes of Indonesia and the western Pacific, and it followed a consistent arithmetical rule: the number of species (birds, reptiles, grasses) approximately doubles with every tenfold increase in area. Take an actual case, the land birds of the world. There is an average of about fifty species on islands of 1,000 square kilometers, and about twice that many, 100 species, on islands of 10,000 square kilometers. In more exact language, the number of species increases by the area-species equation $S = CA^z$, where A is the area and S is the number of species. C is a constant and the exponent z is a second, biologically interesting constant that depends on the group of organisms (birds, reptiles, grasses). The value of z also depends on whether the archipelago is close to source areas, as in the case of the Indonesian islands, or very remote, as with Hawaii and other archipelagoes of the eastern Pacific.[8]

This relationship has been tested and retested by ecologists like Jared Diamond at UCLA dozens of times and for dozens of habitats of varying sizes in the thirty years since it was published, finding in nearly all cases confirmation of the basic formula, but slight variations in the value of z.

Wilson and other ecologists have applied the species-area equation to project the percentage of species that might be lost (driven to extinction) if islands of forest patches are reduced by human developments in the decades ahead. Wilson explains that:

> when an area is reduced, the extinction rate rises and stays above the original background level until the species number has descended from a higher equilibrium to a lower equilibrium. The rule of thumb, to make the result immediately clear, is that when an area is reduced to one tenth of its original size, the number of species eventually drops to one half . . . and is actually close to the number often encountered in nature. . . .
>
> If destruction of the rain forest continues at the present rate to the year 2022, half of the remaining rain forest will be gone. The total extinction of species this will cause will lie somewhere between 10 percent . . . and 22 percent.

How many species do Wilson's projections using island biogeography theory say are going extinct each year? In order to estimate that, we first must estimate how many species there are on Earth, how many are going extinct today relative to the past, and how important are various species in different communities. The Oxford University ecologist (and current science adviser to the British prime minister) Robert M. May has long tried to answer these queries:

> For ill-understood reasons that lie deep in past intellectual fashion, we have made surprisingly little progress toward answering these questions. The date of the stan-

dard edition of Linnaeus's work, which may be taken as beginning the simple factual task of recording the diversity of life on Earth, is 1758. This is a full century after Newton had already given us an analytic and predictive understanding of gravitational laws, based on centuries of recorded information about planetary motions and star catalogs. The legacy symbolized by this lag between Newton and Linnaeus is still with us, and today we know more about (and spend vastly more on) the taxonomy and systematics of stars than about the taxonomy and systematics of the organisms on our planet. We have a better estimate of the number of atoms in the universe—an unimaginable abstraction—than of the number of species of plants and animals currently sharing the Earth with us.[9]

Despite the plaintive cries of ecologists to survey the diversity of nature (and the equally passionate attempts of some politicians and landowners—who do not want to know how many species live on their private holdings—to unfund government efforts to catalog global biological resources), scientists do have reasonable ballpark estimates of biodiversity. Wilson suggests that a conservative estimate for rain forests alone is 10 million species. Given that current destruction rates of forests are at greater than 1 percent per year and applying the species-area formula from the theory of island biogeography, Wilson makes an "optimistic estimate" of 77,000 species lost each year to extinction, twenty-four each day, no fewer than three per hour. Normally, the rate would be about one species lost per one million species present per year. Thus, he asserts, "[h]uman activity has increased extinction between 1,000 and 10,000 times over this level in the rain forest by reduction in area alone. Clearly we are in the midst of one of the great extinction spasms of geological history."

The staggering implications of these ecologists' estimates (even without combining them with other global change disturbances) are plain. Does one species, bent on increasing its numbers and economic standards, have the right to proceed with this unwitting or disinterested planetary-scale slaughter, or rather, should *Homo sapiens* use our evolved capacity to feel and reason to pause on our path, take stock of the potential consequences of our behavior, make ourselves collectively less destructive, and, despite its lack of popular political support, rethink the global-scale value system that puts human numerical and economic growth ahead of nearly all other competing values? Given the enormity of these implications for certain economic interests or world views and our civilization's overall penchant to ascribe higher moral value to human development than to preserving nature, perhaps it shouldn't be surprising that some who are content with business-as-usual, anthropocentric values would labor mightily to discredit both the science and the values of those who fret out loud about a human-induced biological extinction crisis.

DATA-ORIENTED ECONOMISTS VERSUS THEORY-ORIENTED ECOLOGISTS

At the risk of exaggerated stereotyping and the creation of a somewhat false dichotomy, let me nonetheless assert that the professional yin for the ecologists' yang are free-market, conventional economists. The latter usually insist that past data, rather than theory, should be the basis for forecasting. "Data-oriented economists," it was pointed out to me by the University of Geneva natural philosopher Jacques Grinevald, "are in fact a social construction in the modern profession of economics, whose tradition is a very theory-oriented social science." Regardless of their professional evolution, such data-oriented analysts disdain any data-poor theory. Ecologists, worried about causes of changes, also use data, but to develop and test theory; then they use the theory to forecast changes.

However, if some rule derived from past data does not adequately reflect the mechanisms that might be operating in the future, but rather is a simple extrapolation that reflects the dominant mechanisms of the past, then predictive skill for this rule is suspect. This is especially so if the future will be a time in which conditions are likely to be very different from the past—subjected to unprecedented global change disturbances, for example. Then, even a library full of data on past behavior may produce rules with insufficient insight to the unfolding future—unless the rules used to forecast represent the mechanisms that will operate for conditions significantly different than in the past.

All responsible ecologists admit a maddening degree of uncertainty surrounding both the numbers of species and current extinction rates, and are endlessly frustrated by what they perceive to be the lack of public support for census activities to address these data deficiencies. However, all good scientists know that to forecast the future requires a rule of how the system behaves that contains or at least represents causal mechanisms. Most would agree with data-oriented economists that such rules need to be derived and tested on past behavior of the system as much as possible.

The opponents of theory-based approaches, led by the University of Maryland business professor Julian Simon, do not like the implications of the forecast biodiversity crisis if they threaten individual initiative and economic growth. They directly challenge island biogeography as an outrageously flawed theory and suggest that any inferences based on species-area curves are tainted. They point to extensive reduction of the size of the forests in eastern North America during the nineteenth century and say that, using island biogeography theory and the species-area formula, a significant fraction of the region's two-hundred-plus territorial bird species should have—but didn't—become extinct during or since this period.[10]

A BIRD-BRAINED IDEA?
··

The University of Tennessee ecologist Stuart Pimm and Robert Askins of Connecticut College produced a detailed analysis of such critics' attempts to turn island biogeography theory against itself, with the bird extinctions of the U.S. Northeast as their weapon of choice.[11] First of all, Pimm and Askins note, you can't use all 220 species of terrestrial birds that inhabit eastern North America, for only about 160 of these species belong to eastern forests; the remainder live in grasslands, marshes, and other open habitats. The next question is how much forest was cut and what is the proper fraction to use in the species-area formula. They note that deforestation began as a wave progressively moving from the sixteenth century along the east coast inland toward the Mississippi River, leaving very little forested area untouched by ax and saw. However, it would be entirely inappropriate, the ecologists argue, to use, as the critics did, the total amount of forest cleared over a three-hundred-year period in the species-area formula, for that is not how these formulas were derived from past data.

When trees were cut and logging moved westward, not all the deforested land remained cleared. Considerable patches grew back. At the same time, wild birds that may have been driven from cleared patches simply flew to the remaining segments of forests or repopulated the new forest islands that grew back after cleared lands were abandoned, particularly in the twentieth century. In their words, "even during the peak period of deforestation in the late 19th and early 20th century, there were large forest refugia that provided habitat for forest birds. After 1920, the amount of deciduous forest showed a steady increase in the Northeast and the South."

But this is not their prime criticism of the social scientists' failed attempt at playing ecologists, but rather a more fundamental issue of intellectual understanding. The critics who pointed to the small number of extinctions in eastern forests (even if unwittingly) were referring to global extinctions, that is, species lost forever from the world. Ecological *island* theory is not derived for the world, but for particular islands or habitat patches. It predicts *local extinction*, not global extinction. Thus, Pimm and Askins note,

> most of these 160 birds would not become globally extinct even if all of the eastern forests of the United States were cleared. Many species are so wide-ranging that their distribution across say, the relatively undisturbed boreal forests of Canada, makes them invulnerable to forest losses elsewhere. Some of the 160 species are so widely distributed that they would not become globally extinct even if the combined forests of North America, Europe and Asia were clear-cut. The key difference is between global and local extinctions.

To redress this problem, one must apply the theory at the level at which it was derived: local areas. Therefore, the appropriate bird species to study through the species-area formula are those that live *only* in the Northeast—that is, are endemic to the area. Pimm and Askins concluded that adding the four known extinctions leads to an estimate of between 13 and 28 avian species prior to extinction that should be considered as eastern forest endemics. "The possible extinction rates range from 4/13 = 31 percent, if one defines eastern forest birds strictly, to 4/28 = 14 percent," if one adds a dozen or so more species that have 75 percent of their ranges in the eastern forests.

Notice that it is not data itself that is needed but the appropriate kinds of data if scientific formulae are to be correctly applied and tested. The critics used "data," but in a scientifically inappropriate manner. I am reminded of a cartoon showing a road sign on the outskirts of a small town with numbers for the altitude, population, and date of founding all listed above a line. Below the line is the total of the three disconnected numbers. An obvious cartoon absurdity was also performed with real data by critics who apparently didn't understand the hypothesis being tested. It requires some degree of scientific knowledge even to estimate something as seemingly straightforward as how many birds are resident only in the U.S. Northeast. These details were obviously overlooked by the critics in their rush to apply "data," however inappropriate, to "test" the theory whose conclusions they didn't like.

In fact, the species-area formula actually slightly underestimates the number of birds that should have gone extinct in New England, rather than overestimates it by a factor of six, as claimed by the nonbiologist critics who misapplied the fundamental theory. Furthermore, the ecologists conclude, other factors have also been associated with bird extinctions, such as hunting. Whereas this may have been exacerbated owing to habitat fragmentation that concentrated the birds, making it easier for the hunters to find them, it would help to explain why the species-area formula actually underpredicted the number of global extinctions relative to those that actually occurred.

Finally, what can be learned about extinctions in tropical forests from the flap over the application of island biogeography theory to the U.S. Northeast bird extinctions? Pimm and Askins address this in their concluding remarks, noting that many tropical areas are not only rich in species, but are unusually rich in endemic species. Pimm, who once took Terry Root and me to see a small, beautiful, critically endangered Hawaiian bird (the akiapola'au) high on the slopes of the Mauna Kea volcano on the Big Island of Hawaii, one of its last habitat patches, said, "Look at it well; it will probably be gone before either of you are."

He noted that Hawaii once hosted 135 species of terrestrial birds, that all were endemic, and that all but 11 are now either extinct or critically endangered. As in the Northeast example, habitat losses applied to the species-area formula predict far fewer than that number of species, because this is only one factor, and many other disturbances are also responsible for the extinctions. "To predict global extinction, one needs an understanding of patterns of endemism," Pimm and Askins wrote. "Eastern North America lost few species of birds despite heavy forest losses because it has so few endemics. The proportion of endemic losses is higher than expected. We find this conclusion only supports those who express concern about deforestation and species loss throughout the world." They are especially concerned about tropical deforestation because of the high degree of endemism in tropical forests.

However, as this example shows clearly, there is so much complexity and missing information that there is still great uncertainty in making extinction forecasts, whether by species-area formulas or other cloudy crystal balls. Newspaper editorials by critics who don't understand how to apply the theory certainly do little to clear the glass.

SYNERGISM AND UNCERTAINTIES

What, then, is the message of all this controversy over biodiversity loss? I think it is safe to say that simplistic assertions that dismiss concern for species loss because of a current lack of data on total numbers of species or extinction rates, while ignoring well-confirmed ideas that can produce credible but gross estimates, are, at best, poor scholarship and, at worst, public policy polemics. To deny that undiscovered species are going extinct in hacked-down tropical forests because no trained ecologist is there to take a census is analogous to denying that forest fires are being started in dry, remote, unobserved wilderness when lightning storms are present.

Data *by itself* has no predictive value unless the formulae it generates or validates reflect the causal processes of change.

The extinction rate predictions of the island biogeographers are based on one causal factor: habitat loss. It is my personal view that by far the most serious environmental problems of the twenty-first century will not simply be habitat loss or ozone depletion or chemical pollution or exotic species invasions or climatic change by themselves, but rather the synergism of these factors.[12] As Stuart Pimm noted, cut forests may not lead to more than local extinctions, for the displaced birds can fly to neighboring patches. But as those patches shrink and become few and far between, it is more difficult for the remaining species crowded in these refuges to migrate if the climate changes out from under them, particularly if that change is a factor of 10 or more faster than the sustained globally average rates of change such species had to cope with since the last ice age. Imagine the difficulty the forest species of ten thousand years ago would have experienced if, in their individual migrations to the north, they also had to cross the factories, farms, freeways, and urban sprawl of twenty-first-century Earth, and, on top of all that, also experienced climate change twenty times more rapid.

If one bird species vacated old territory, might not an outbreak of pests occur in their wake? On the other hand, perhaps those migrating birds could reduce the pests in another zone to the north. These are the kinds of speculations over biotic community functioning that frustrate ecologists and conservation biologists, for it is difficult enough to explain the distribution and abundances of species today, life that has had ten thousand years to settle into their current ranges without overwhelming amounts of human disturbance until recently. Now, the experiments being performed on Laboratory Earth are demanding of the scientific establishment explanations of the detailed responses of species and communities subjected to multiple disturbances at potentially unprecedented rates. What honest scientist can proclaim certain knowledge of such a questionable future? One wonders what honest critic can chide the scientists for admitting that uncertainty and then use that uncertainty as an excuse to delay action that could lower the risks. That may be good business or political practice for some, but for me it constitutes planetary gambling with the biological riches of Earth.

I think you can see by this battle of the birds just how complex Earth systems science can be, especially when it is brought down to the level of your local forest or songbirds. The sources of data and their theoretical inferences can be as diverse as the populations to be studied. It is possible to select data and theory in a limited context to support virtually any conclusion. The wise analyst has to look at the whole context and decide whether he or she thinks the preponderance of evidence is compelling. If, like Simon and some data-oriented economists, you find the conclusions not to your liking, it is entirely possible to attack them at the foundations by pointing out contradictory or missing elements of data or theory.[13] There will always be missing data and some elements of weakness in all theories. If you are, like me, terribly concerned about the risks of standing by until all consequences are conclusively demonstrated—if it ever will be conclusive—you can find in these examples from Earth systems science sufficient reason to urge earnest action. I freely admit this requires taking a leap of faith from state-of-the-art, often fuzzy, knowledge to concrete actions. But what if the ecologists are right?

IS BIODIVERSITY WORTH PROTECTING?

I believe that local habitats, like a nearby forest, are part of a continuum of interlinked ecosystems that range from microbial communities at the rootlet of a sapling to the global biogeochemical cycles and the climate system. Diversity—of nature and culture—and sustainability are the values I most embrace, but I recognize that defining trade-offs in specific cases is not amenable to a "scientific" formula. What is needed is a process of management in which people knowledgeable in the terms of the debate have enough mutual respect (and rules of engagement) to work for solutions that do not entirely ignore the interests of anyone across a wide spectrum of actors. Such solutions must also recognize that humans are but one species and their rate of growth and material progress simply is one value, not the supreme purpose, of planetary management.

I use the term *planetary* self-consciously, since the interconnectedness of ecosystems transcends the discrete scales that characterize private holdings, nation-states, or disciplinary practices. Environmental management needs to be at the scale of the managed system, but most existing organizations and institutions are set into physical and legal boundaries not always in sync with the scale of the problems. There may be a need for a continuum of management models to parallel the interconnected human and natural systems that fall under the rubric of environmental sustainability. To correct these mismatches requires a willingness to cede some local or national sovereignty to management coalitions that better fit the configuration of the actual systems under consideration. Finally, I hope that the culture of "optimizers" with a bagful of analytic decision tools will recognize that the details of very complex systems are not always predictable, even if likelihood of change is. Thus, we should consider supplementing the paradigm of optimization of economic efficiency with one of risk aversion (the so-called precautionary principle) that seeks to preserve a major measure of diversity for culture and nature.[14] "We first" could well eventually result in "nature last."

CHAPTER 6

INTEGRATED ASSESSMENTS OF POLICY OPTIONS

I can summarize in two sentences what has taken five chapters on Earth systems science to demonstrate: There is a wide range of scientific opinion on precisely what might be the impacts of climate and other global changes that could result from a number of human activities. These outcomes range from beneficial effects on plant growth from extra CO_2 in the air to potentially catastrophic regional climatic impacts on agriculture, water supplies, coastlines, health, and species. In this chapter, I will show why current scenarios of the future should certainly not be taken literally, but their implications do deserve to be taken seriously.

OPTIMIZING EFFICIENCY

While there is much disagreement over the details, if one were to sample a large fraction of the knowledgeable scientific community, I am confident that one would find that the bulk of such experts consider (1) it to be a pretty good bet—a coin flip at least—that substantial global change impacts will unfold, (2) that there's perhaps only a 10 to 20 percent chance that either net beneficial to negligible changes will result, and (3) there is also a 10 to 20 percent chance that there will be widespread catastrophic outcomes from the human activities creating global change. These estimates of the consequences of global changes are called impact assessments, and represent the potential costs, monetary and non-monetary, to society and nature if the global change experiment now under way continues and is intensified.[1]

Are You a Rational Actor?

Before what economists and political scientists call a "rational actor" would demand action to avert such global changes, such a mythical being would first want to know what the costs of mitigating the global change impacts might be (for example, the cost to the economy of taxing carbon dioxide emissions). This rational being would then want to spend only as much in averting climate change as the impact assessments suggest the change would cost if it occurred unabated. This is called an economically efficient optimization policy.

Is an Efficient, Free Market an Oxymoron?

What do economists mean by *efficient*? A paramount faith of most economists is that a "free market" is the most effective way to achieve economic well-being. To true believers, allowing unfettered functioning of the market is the top priority of government, not health regulation or environmental protection if these constrain free markets. By *free market* economists mean that if people are free to spend their money as they see fit, since it is in their interests to minimize costs and maximize benefits, they will, collectively and over time, find the best way to be *efficient*—that is, find the least-cost and highest-return solutions to the business of life (or at

least for the life of business). This is what founding economist Adam Smith meant when he said that the economic
system is best guided by the invisible hand of the market
rather than the long arms of government officials. In the several centuries since, a host of economists, business leaders,
and politicians have repeated that catechism at the first sign
of political trouble—usually in the form of proposals for government regulations of entrepreneurial activities.

One of the big debates today is over whether markets are
as efficient as billed, even if free (we'll deal with the "free"
part later on). An environmentally relevant example is the
debate among technologists and some environmentalists
over why most of our houses, stores, factories, and offices
do not use the most energy-efficient bulbs or windows, for
example. Engineers argue that the extra energy used from
this inefficiency costs more over time than the investment
needed to upgrade the efficiency of our buildings. Believing
economists typically reply that if all costs are accounted for,
people are not changing their behavior because that must
not have been an overall cost-effective upgrade. This rests
on the assumption that people efficiently serve their economic self-interest—it would self-evidently be irrational to
do otherwise! Technologists often reply that, despite two
hundred years of "efficient markets" rhetoric, free markets
are not even economically (let alone energy) efficient. For
instance, take the average house in the United States and
look at the windows, the insulation, and probably the car
outside. Or look into the machinery that's in the typical
manufacturing company. Chances are that none of these are
efficient in terms of energy usage, compared to the best
available products at the state of the art. But to measure economic (as opposed to energy) efficiency, we need to ask: Is it
more expensive to replace these energy-inefficient products
than it is to pay the extra energy bill (never mind for the
moment the fact that using that extra fuel produces extra
pollutants that can disturb the environment, which itself

has some costs to society as a whole that economists call "externalities"). Even leaving out the externalities, there are many critics who assert that the reason we haven't installed all the energy-efficient products available is that the free-market system doesn't work as economically efficiently as claimed. This is called a "market failure." Economists usually admit that the free market isn't a perfectly efficient economy, but blame most of that on government interference, not so much on investor or consumer errors. It takes clear proof to get a confession of market failure from many free marketeers.

I witnessed such debates between economists and engineers on a 1991 U.S. National Research Council assessment of the policy implications of global warming.[2] Our charge was to advise the U.S. government of the potential seriousness of the issue, potential policy responses, and their costs and benefits. That set the stage for the engineer/economist debate. It was often entertaining to listen to them go at each other: "It's incredible that we could cut CO_2 emissions by 10 to 40 percent and make money doing it," a technologist might say, "yet so-called free markets are stuffed with outmoded, energy-inefficient products." "No, the reason company X isn't running out to buy all the latest gadgets," an economist could retort, "is because it isn't economically efficient for the CEO of a company to be saving the firm 10 percent by spending months figuring out how to improve the energy efficiency of a plant. That CEO should be spending his time reading the *Wall Street Journal* and getting a higher than 10 percent return on investment by learning about new business opportunities." Some true believers will struggle mightily, I mused to myself, to find a way to argue that the market is close to efficient. "I agree that any CEO spending his or her time running around the company trying to find what bulbs are energy-inefficient is not a very good CEO," I chimed in, "but wouldn't you expect a good executive to hire a technologist at a much cheaper salary to do the job and give him or her the authority to make the company more energy- and cost-efficient?"

In any case, this is the sort of example that motivates a well-traveled economist joke. There is a Nobel Prize–winning economist who enjoys walking with his granddaughter in the park. One day, she skips away toward a park bench. "Wait, Grandpa," she says, "I want to go over to that bench and get something." "Why, dear, what are you going for?" "I see a twenty-dollar bill under the bench," she replies excitedly. "No, there isn't," the economist counsels authoritatively while looking elsewhere, "because if there were, someone would have taken it already." Even a loose twenty-dollar bill is an unthinkable market failure! (Imagine my pleasure last summer when my teenage daughter found a twenty-schilling bill on the ground in a crowded shopping street in central Vienna. I instantly insisted she recount the event to the economist for whose child she baby-sits!)

Is the market efficient? That's not a dispassionate debate in this community. Market failures occur if people are unaware of cost-efficient opportunities (like energy-efficient new products) or if there are hidden subsidies (for example, a law that limits the liability of power companies if there were a nuclear power plant accident keeps the price of nuclear electricity lower, but does not let the market work because it subsidizes one form of energy supply at the expense of cheaper competitors). That is not economically efficient. For a market to be efficient, it should maximize benefits and minimize costs to the society as a whole. But economists have long struggled with the question, Costs and benefits to whom? What if the benefits accrue to the investors, but the costs fall, often invisibly, on others external to the market? That is a question of equity rather than efficiency. Even if the market were efficient by standard accounts, which is normally calculated in terms of recorded transactions like income and expenses, there is still the question of how the price of energy, for example, can be at a "fair market price" if the cost of energy doesn't include the fact that out of the tailpipe comes an effluent that damages someone else's lungs, alters the climate, or does other things with potentially detrimental effects for segments of society.

If you believe in a true free market that is efficient, you have to include at least something in the price of doing business to account for social costs external to standard accounting. This is not simply a technical question of finding the right external cost. "Internalizing externalities," as economists refer to this issue, is a cultural and political question as well. For example, how are we going to deal with the price of energy, knowing that pollution causes a degradation of common property, but that an energy price increase may temporarily hurt the economy and poor people especially—pitting equity versus efficiency versus environmental protection? Another cultural obstacle is that individual decisions affect global "property," but no one owns the global environment—it is a so-called commons. Currently, unless a regulation or user fee is in force, there are few incentives other than moral persuasion to induce individuals to reduce their impact on commons. Nearly all modern economists acknowledge this to be a potentially serious market failure. The problem has even given rise to a splinter wing of economists (and natural scientists too) who created the field of "ecological economics."[3]

Returning to the issue of efficient optimization policy, how can such an optimization be calculated? Global change scientists use a kind of modeling called integrated assessment in an attempt to merge economy and ecology in a cost-benefit framework. Some boosters of analytic methods such as cost-benefit analysis repeatedly insist that real-world decisions should be subservient to the answers from these rational tools and that anything else is less than objective, or even irrational. That belief, no matter how earnestly held, is not a scientific statement, but rather a cultural bias. I'm not arguing that all decision making should be irrational or ignore the insights of such analytical methods. On the contrary, I devote a significant part of my professional life trying to understand such methods, since they can help to inform us on those aspects of economic issues that are measurable in commensurate units of currency. It would indeed be irrational to ignore them, and the fact that so few people understand the circumstances in which economic tools apply to policy processes is a real threat to an efficiently (politically, that is) functioning democracy. But the steadfast belief of some that market efficiency and its analytic toy, cost-benefit analysis, should be the *sole* basis for policy making has led to significant conflicts with ecologists and many economists as well.

As we'll see, integrated assessment falls short on several grounds because of the technical difficulty in evaluating credibly the whole range of system costs and benefits of our activities and because, ultimately, the single unit of comparison is monetary—usually the dollar. Many things real people (as opposed to rational actors) value are not easily monetized (freedom, nature, love, security, or the value of a human life). Integrated assessment is, however, the best science can do, and is far better, I believe, than the piecemeal approaches, infomercials and other advertisements, press releases faxed to media and government offices, and emotionally appealing but often unrepresentative stories used (and abused) in political debates.

Since we've been focusing on global climate change, let's use that example for our integrated assessment of global change case study. To discern impacts, we need to begin with an estimate of how much greenhouse gas or sulfur dioxide people will produce over the next hundred years in order to assess the long-term consequences on atmospheric concentrations. That means projecting what technologies will be in place, how much of them each of us will have access to, and how many people there will be on the Earth (the I = PAT mentioned in the introduction). Remember throughout this discussion that the full cost of things is seldom what it seems. The price of a lump of coal isn't simply extraction, storage, and transport, but health consequences of mining and burning, as well as the whole range of potential environmental alterations from the production and use of energy. The cost of a car isn't simply the materials, labor, and profit, but should also include disposal costs and tailpipe emissions. These costs are normally external to standard economic accounts, what I've already defined as externalities. But they are real costs to society nonetheless, even if conveniently omitted from conventional "rational" political or economic discourse.

Scenarios

The United Nations puts out population projections, typically including a high, medium, and low *scenario*—the absolutely favorite word of integrated assessors—of future population. There is an incredible difference between the projected population in 2100 in the low and high scenarios—about 5 billion versus 20 billion. The primary difference in these three scenarios is how soon various now-developing countries reach "replacement level" of fertility—each couple having, on average, about two children. The low scenario is considered improbable by nearly all analysts, since it assumes below-replacement fertility in the next century. Most futurists believe that reality will fall between the medium and the high scenarios (10 to 15 billion, perhaps), depending on how rapidly fertility rates decline in various countries.

It is mind-boggling to realize that the world could have 10 billion more people in it in 2100 if every other factor is constant except that there is a delay of a few decades in achieving replacement fertility in the developing world. Those who argue that family planning and other programs designed to reduce birth rates will make only a marginal difference often point to the year 2025 for their conclusions,

since there is relatively little difference among the three scenarios that soon. However, there is a problem in the age
structure (the percentage of the population in each age
group), which sees most high-population growth rate countries having a disproportionate fraction of their citizens still
below reproductive age. Thus, even if replacement fertility
were reached tomorrow, that is, each of these potential couples has only a two-child family, the population must continue to grow—a concept known as the momentum of population. The reason is that zero population growth comes only
after the disproportionate percentage of young people have
had their children. If all couples now and in the future
somehow agreed to practice replacement fertility tomorrow,
in seventy-five years the 1990s population of some 5.5 billion would still have grown, owing to momentum of population, to about 8 billion.

Therefore, the sooner replacement fertility can be
achieved, the dramatically lower the ultimate population
size. This explains in part why the Chinese opted for the
draconian and controversial policy of one-child families in
order to prevent the momentum of population from seeing
China double its size before the end of the next century.
Depending on your ethical views or which side of the opti-
mist/pessimist, economist/ecologist paradigm gulf you're
on, you might see either the coercive policy or the advent of
a doubled population size as the greater evil. That value
judgment aside, let's consider it within the integrated assessment process.

Once we choose a scenario for population growth, we
need to determine how much consumption each person will
have—in other words, our affluence level as traditionally
defined.[4] Current per capita income in lesser developed
countries averages somewhere around $1,000 per person.
This compares to about $10,000 average per person in the
richer world, the more developed countries. Clearly, equity
and politics demand development in the less developed

countries, and every official document written by any national government or international agency, such as the UN, assumes dramatic economic growth rates in the per capita conditions of the world's poorest 80 percent. The analysts, along with a majority of political economists, believing that greed will always be part of our personal psychology, assume that citizens of the more developed countries will not sit idly by in the twenty-first century while the poor catch up to them in per capita consumption. Rather, there is a projected increase in the personal income in the more developed countries as well. A typical projection is for a 400 percent growth in world per capita consumption by A.D. 2100, with perhaps an 800 percent growth for the less developed nations. The job of the nonadvocate analyst is to run a whole host of plausible cases and ask the question: What if scenario A versus B versus C were to occur? In that spirit, let's simply adopt these typical populations and affluence projections as plausible.

The next chain in the logic of integrated assessment is critical to environmental concerns, for it involves what technologies will be used to achieve this severalfold growth in per capita consumption for a world of 12 billion people. First is the all-important term "energy intensity." This refers to the amount of energy necessary to produce a unit of economic product. In the richer countries, the amount of energy needed to produce a unit of Gross National Product—the standard measure of economic well-being—has come steadily downward in the past several decades. The average rate of improvement (less energy required for each unit of GNP) was about 1 percent per year over the last thirty to forty years. It took a jump upward in the decade after the 1973 OPEC energy price rises, as price-induced rationing, innovation, and energy efficiency accelerated at a rapid rate because there was a price incentive to do so.

It is a major controversy in economics to guess what this rate of improvement in energy intensity should be in the future. Of course, as the OPEC embargo lesson showed, the price of conventional energy (fossil fuels) should also be part of the equation. Nonetheless, very few such integrated assessment models make this obvious assumption yet. Rather, the assumption typically used is that improved, more energy-efficient products will be developed at some constant rate over time because it takes people time both to create and to purchase such products, and somehow the price of energy is irrelevant. My Stanford colleague, the economist Larry Goulder, and I have shown, using his energy-economy model, that this standard assumption of price irrelevancy may bias the presumed costs of carbon taxes upward, and thus future-generation integrated assessments must take into account how the increased price of conventional energy from policies to reduce greenhouse gases could induce technological improvements that either accelerate energy efficiency or reduce the price of nonconventional energy (for example, solar).[5]

In some less developed countries (LDCs), the energy intensity has actually been getting worse, not better, in the last decade, but nearly all analysts expect this ominous trend to reverse as LDC economies heat up, capital is generated, and more efficient products already available in the marketplace can be purchased and installed.

As we saw over the battle of the birds, predicting what will happen in the future based on the data we have from the past is tricky. The forecast of eventually improved LDC energy intensity is based on the Western experience, which suggests that when economies begin to develop rapidly they will use any cheap and dirty means to power their industrialization—typically coal, as is the plan for China and India today, just as it was in the Industrial Revolution in the United States and England over a century ago.

Again, based on Western experience, it is considered likely—but not certain—that in a decade or two it will be economically cost-effective for rapidly growing, developing country economies to implement more energy-efficient, less polluting technologies than they now do in the early phases of their development process. Since none of this is certain, it is an area where alternative scenarios that reflect different policies and practices can provide the integrated assessor an opportunity to make a contribution to the policy-making process. By showing the relative importance of different policies that affect energy-intensity scenarios, we can, in the end, show very different impacts on the amount of carbon dioxide or sulfur dioxide emissions, and therefore potential climate change and unhealthy air pollution, associated with alternative development policies.

The implications of alternative projections for population, affluence, and technology can also be assessed—what is called "sensitivity analysis." Such policy analysis is one of the virtues of integrated assessment tools and is why the technique has been embraced by many national governments as a rationalist's way to try to pull together the many disparate parts of the global problem of climatic change.[6]

Energy, per se, is not the primary problem for the global climate; carbon emissions is. If biomass, solar, or nuclear were the prime energy sources, then the amount of carbon emitted per unit of energy would be much smaller than if coal or oil, or synfuels based on the carbon in coal, would power society. Or, in the interim, natural gas can be used since it pollutes much less than the other fossil fuels. Therefore, an additional factor is needed in our equation that calculates carbon dioxide emission from population, affluence, and technology trends, and that factor is the so-called carbon intensity. This is the amount of carbon dioxide emitted per unit of energy produced. Here is another example where sensitivity analysis can be performed by integrated assessments in terms of the effect on CO_2 buildup from a natural gas rather than a coal-based future economy. (This would reduce CO_2 emissions by more than a factor of two.)

Guessing with Experts

It is absurd to think that any current analyst could know for sure which particular scenario will emerge. In fact, strictly speaking, the probability of any one of them is zero. That is, it is virtually certain that the future will not unfold precisely along any one curve that we draw. The purpose of integrated assessment is not to provide an exact forecast of what will happen, but rather to show the differential consequences of alternative assumptions. Regardless of the value system of the analyst, at least the analysis process makes explicit the logical consequences of alternative choices.

If any one choice has a zero probability of being exactly true, how can we present the results in a meaningful way in the real world?

One solution has been tried by the Environmental Protection Agency analyst James Titus. He and colleagues worked on a small piece of the integrated assessment mosaic—the impact of various CO_2 emissions scenarios on sea level rise. He has long been interested in the question of sea level rise associated with possible ice cap melting or with warming the oceans and resultant thermal expansion of the waters. Either process could cause the waves to lap a little bit higher on our shores. (Sea levels rose between 10 and 25 centimeters in the twentieth century and typically are projected to rise from 0 to 120 centimeters more in the next hundred years, with a "best guess" of about 50 centimeters.[7]) Studies of the economic values of coastlines suggest that sea level rise could be a hundreds-of-billions-of-dollars externality in the future, and thus estimating the probability of its occurrence is a very significant integrated assessment activity.

Titus, recognizing that no one scenario could be credible, tried to examine a family of plausible outcomes and give each of them a probability as objectively as possible. To do this, he relied on the subjective best guesses of several dozen experts as to how much CO_2 will be emitted in the future, how nature will dispose of it through the natural carbon cycle, how CO_2 will translate itself into climatic change, and how that climate change will affect the ice masses at the poles and the temperature distribution in the ocean—which together determine sea level changes.

Titus and colleagues—including teams of experts of all persuasions on the issue—calculated the final product of their impact assessment as a statistical distribution of future sea level rise, ranging from slightly negative values (that is, a sea level drop) as a low-probability outcome, to a meter or more rise, also with a low probability (see figure 6.1).[8] The midpoint of the probability distribution is something like a half-meter sea level rise by the end of the next century. Titus cautions not to take the numbers literally, but I believe the overall structure of the results in figure 6.1 does portray the issue fairly and should be taken seriously.

Since the EPA analysis stopped there, this is by no means a complete assessment. In order to take integrated assessment to its logical conclusion, we need to ask what the economic costs of various control strategies might be and how the costs of abatement compare to the economic or environmental losses from sea level rises. That means putting a value—a dollar value, of course—on climate change, coastal wetlands, fisheries, environmental refugees, and so on. Hadi Dowlatabadi at Carnegie Mellon University leads a team of integrated assessors who, like Titus, combined a wide range of scenarios of climatic changes and impacts but, unlike the EPA studies, added a wide range of abatement cost estimates into the mix. Their integrated assessment was presented in statistical form as a probability that investments in CO_2 emissions controls would either cost more than the losses from averted climate change or less.[9] Since his results do not include estimates for all conceivable costs (such as the political consequences of persons displaced from coastal flooding), the Carnegie Mellon group offered its results only as illustrative of the capability of integrated assessment tech-

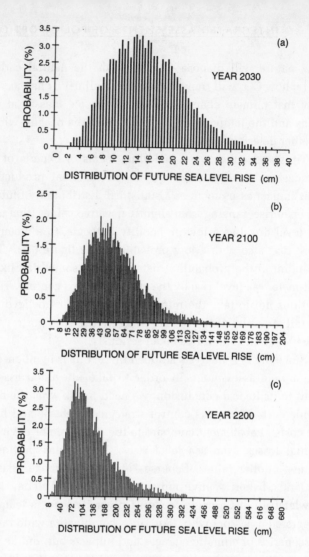

FIGURE 6.1

Rather than pretend we can know precisely what will happen to climate in the twenty-first century, it is more appropriate to express future scenarios as a probability distribution. Here, James Titus and V. Narayanan have compounded a range of possible CO_2 emission estimates with a range of possible CO_2 concentrations and used the latter to calculate a range of possible climatic changes (sea level rise, in this case). While the probabilities calculated for each specific outcome cannot be taken as quantitatively reliable, the overall shape of the graphs and magnitudes of the changes are representative of state-of-the-art assessments.

niques. Its numerical results have meaning only after the range of physical, biological, and social outcomes and their costs and benefits have been quantified—a Herculean task.

Similar studies have been made by a Dutch government effort to produce integrated assessments for policy makers.[10] Jan Rotmans, who heads one of their efforts, likes to point out that such modeling of complex physical, biological, and social factors cannot produce credible "answers" to current policy dilemmas, but can provide "insights" to policy makers that will put decision making on a firmer factual basis. Understanding the strengths and weaknesses of any complex analytic tool is essential to rational policy making, even if quantifying the costs and benefits of specific activities is controversial.

Yale University's William Nordhaus has taken heroic steps to put the climatic change policy debate into an optimizing framework. He is an economist who has long acknowledged that an efficient economy must internalize externalities (in other words, find the full social costs of our activities, not just the direct cost reflected in conventional "free-market" prices). He tried to quantify this external damage from climate change and then tried to balance it against the costs to the global economy of policies designed to reduce CO_2 emissions. His optimized solution was a growing carbon tax, designed to internalize the externality of damage to the climate by increasing the price of fuels in proportion to how much carbon they emit, thereby providing an incentive for society to use less of these fuels.

Nordhaus imposed carbon-tax scenarios ranging from a few dollars per ton to hundreds of dollars per ton—the latter which would effectively eliminate coal from the world economy. He showed that, in the context of his model and its assumptions, these carbon-emission fees would cost the world economy anywhere from less than 1 percent loss in Gross National Product to a several percent loss by the year 2100. The efficient, optimized solution from classical eco-

nomic cost-benefit analysis is that carbon taxes should be
levied sufficient to reduce the world economy as much as it is
worth to avert climate change. He *assumed* that the impacts of
climate change were equivalent to a loss of about 1 percent of
GNP.[11] This led to an "optimized" carbon tax beginning at ten
dollars or so per ton of carbon dioxide emitted. In the context
of his modeling exercise, this would avert only a few tenths of
a degree of global warming by the year 2100, a very small fraction of the 4°C warming his model projected.

How did Nordhaus arrive at a climate damage being
about 1 percent of GNP? He assumed that agriculture was
the economic market sector most vulnerable to climate
change. For decades, agronomists had calculated potential
changes to crop yields from various climate change scenarios, suggesting that some regions now too hot would sustain
heavy losses from warming whereas others, now too cold,
could gain. But the agrometeorologist Norman Rosenberg
pointed out that such agricultural impact studies implicitly
involved the "dumb farmer assumption." That is, they
neglected the notion that farmers can adapt to changing market, technology, and climatic conditions. Economists like
Nordhaus believe that such adaptations will dramatically
reduce the climate impact costs to market sectors like farm-
ing, transportation, coastal protection, and energy use. Ecologists, however, dispute this optimism as complacent since
it neglects such real-world problems as people's resistance
to trying unfamiliar practices, problems with new technologies, unexpected pest outbreaks, and the high degree of variability of weather, which will mask the slowly evolving
human-induced climatic signal and discourage farmers from
risking unfamiliar adaptation strategies.

I was recently engaged in debate with one optimistic agricultural economist who asserted that modern farmers could
overcome virtually *any* plausible climatic change scenario. I
countered that he conceived of these farmers as all plugged
into the electronic superhighway in real time, aware of the

probability distributions of integrated assessments, and financially and intellectually capable of instant response to a bewildering array of changing pest, crop, weather, technology, policy, and long-term climatic conditions. "You have replaced the unrealistic 'dumb farmer' assumption of the past," I went on, "with the equally unrealististic genius farmer." Real farmers are likely to fall somewhere in between. And, especially in developing countries, problems with agricultural pests, extreme weather, and lack of capital to invest in adequate adaptation strategies will be a serious impediment to reducing climatic impacts on agriculture for a long time, even for a genius farmer.[12]

Winners and Losers

There is an additional problem with conventional economic analyses of the potential costs of climate change on market sectors like agriculture or coastline change: the prospect of winners and losers.[13] The field of "welfare economics" calculates net changes to overall (not individual) economic welfare of various activities and events, and such costs and benefits from the climate change estimates used in integrated assessments. Thus, if Iowa farmers were to lose $1 billion from reduced corn yields associated with warming, but Minnesota farmers simultaneously gained $1 billion from longer growing seasons, then net U.S. economic welfare change would be zero. Politically, I doubt the situation would be regarded as neutral, as most people believe that equity considerations may require compensation from the winners to the losers. This entire question of "redistribution costs" is difficult, and now is part of the climatic impact assessment debate, but certainly will not go unnoticed in the political arena.

In this connection, the Yale University forestry economist Rob Mendelsohn has tried to estimate the costs and benefits (impacts) of typical global warming scenarios in the United States using what are known as "hedonic" methods. In brief, rather than account explicitly for complex physical, biological, and social processes that determine the profitability of agriculture or forestry, this method simply compares these economic activities in warm places like the Southeast and colder places like the Northeast. This provides a proxy for how temperature changes might affect these segments of the economy. The method is controversial, since natural scientists dispute that the difference between business as usual in northern climates or southern climates can act as a proxy for time-evolving or transient changes in temperature, and other variables, to say nothing about surprises. In essence, these methods assume a perfect substitutability for changes at one place over time with changes across space at the same time—a debatable assumption. But the point here is not to dispute the conclusions, only to report them in the context of the winners and losers issue. Mendelsohn's prime finding with this method is that more heat will make already hot places poorer and currently cold places richer. Countries like Canada win and India lose—a sort of "neo-climatic determinism" reminiscent of that espoused at Yale by Ellsworth Huntington ninety years ago. Mendelsohn wisely acknowledges that even if his conclusions were that the rich countries with big economies and colder locations win more economically than poorer countries that typically are in hot climates lose, this is not a conflict-free scenario, particularly since the standard economic evaluation for the "value" of a statistical human life in rich countries is ten or more times greater than for citizens of poor countries. However, international redistribution of climatic resources currently has no clear legal or governmental authority since nation-states dominate international relations. Climatic change is a global commons issue that will raise tough management dilemmas.

Subjective Experts

I, and a number of other people, debated Bill Nordhaus in a series of letters published in the journal *Science*, arguing that a flat 1 percent loss of GNP from climate change is way too low (see page 168, note 11). Losses could be very high—particularly if there are climate surprises. To his credit, Nordhaus reacted to his critics by extending his study to include a broad spectrum of opinions on the value of damages from hypothesized global warming. He soon published an alternative approach to evaluating the external cost of climate change.[14] Instead of guessing the cost of climate change himself, he recognized that he had included nothing for so-called nonmarket sectors, such as the value of lost species, the value of lost wetlands from sea level rise, the costs from conflicts that might be induced by the creation of environmental refugees or any of the other nonmarket amenities. Since these defy simple quantitative treatment, he took an alternative approach—he sampled the opinions of a wide range of experts who have looked at climatic impacts and asked them to provide their subjective opinions (that is, range of best guesses) on what they thought the costs to the economy would be from several climate-warming scenarios.

The numbers themselves are less interesting than the cultural divide that his study revealed. Nordhaus sampled the opinions of classical economists, environmental economists, atmospheric scientists, and ecologists. The most striking difference in the study was that almost all the conventional economists considered even a radical scenario in which a 6°C warming would unfold by the end of the next century (a scenario I would label as catastrophic, but improbable—maybe only a 10 percent chance of occurring) as not very catastrophic economically. Most conventional economists still thought even this gargantuan climatic change—equivalent to the scale of change from an ice age to an interglacial epoch in a hundred years, rather than thousands of years—

would have only a few percent impact on the world economy. In essence, they accept the paradigm that society is almost independent of nature. In their opinion, most natural services associated with current climate can be substituted with relatively little harm to the economy.

On the other hand, the group Nordhaus labeled as natural scientists thought the damages to the economy from the severe climate change scenario would range from no less than several percent lost up to 100 percent—the latter respondent assigned a 10 percent chance of the virtual destruction of civilization! Nordhaus suggested that the ones who know the most about the economy are optimistic. I countered with the obvious retort that the ones who know the most about the environment are worried.

To paraphrase and caricature the debate, one could characterize the conventional economists as saying that virtually everything we do or virtually every service provided by animals, vegetables, or minerals is substitutable—at a price, of course. If the price gets high enough, someone will invent a way of doing it differently from the way we—or nature—do it now.

If industry runs out of cheap copper, someone will find a substitute material. If construction runs out of lumber, there are cement blocks. Ecologists typically disagree, arguing that natural ecosystem services such as genetic diversity to provide food plants, wetlands for filtering of wastes, or forests for flood control and maintaining proper levels of greenhouse gases to preserve the accustomed climate are not all substitutable at any practical price. Therefore, society shouldn't just bank on the stereotypical faith of economists that we'll somehow find an affordable substitute for each crisis and buy our way out of global scale disturbances, just as we often have before at smaller scales.

The standoff in belief systems conjures up the joke about the economist and the ecologist, good friends who, while hiking together, were arguing about the sustainability of nature and the substitutability of human inventions. This unlucky time, they were walking up on a high ledge, in full and typical animated debate. All of a sudden a great gust knocked one off and, as his friend tried to catch him, they both tumbled to a free-fall. As they sped ominously downward, the ecologist yelled, "Well, I guess we'll never resolve this debate in our lifetime." The economist, ignoring him, yelled, "20, 80, 160, 430."

Finally, with the end approaching, the ecologist shouted in bewilderment, "What are you doing?"

"When the price gets high enough," his true-believing partner retorted, "someone will sell us a parachute!"

Robert Frosch, a former administrator of NASA and then vice president of research at General Motors, went so far as to calculate how many battleship cannons aimed skyward and loaded with dust bombs targeted on the stratosphere it would take to reflect away enough sunlight to offset warming from a CO_2 doubling. The annual costs of this geoengineering project were in the tens of billions, but less than the cost of fuel taxes, he argued.[15]

When an ecologist worries about the limited capacity of ecosystems or species to withstand all human disturbances and thus counsels humanity to lighten the load—even by economically costly adjustments—economists rightly remind us that there are limited human, technological, and economic resources available for all laudable purposes, and we simply can't afford to hedge against every potential ecological impact. We can't afford (or even know how) to replace all natural ecosystem functions like pest control, let alone to replace species we've driven to extinction. Once they're gone, they're gone, ecologists rightly remind us; it is not good stewardship or economics to mortgage our environmental future and leave the burden of finding solutions to our posterity. But we're leaving them more wealth to cope with these burdens, the economists retort. Finding the balance of values across this cultural dichotomy is what the political process is supposed to do. And that process works only to the extent that we can get our values placed into the decision-making cauldron. That is especially hard to do when it is so easy to get confused by an exaggerated and baffling debate.[16] Earth systems science knowledge, including integrated assessments, can help demystify the debate.

Half Empty or Half Full?

The Rockefeller University economist and technology analyst Jesse Ausubel has long thought about the problem of whether technology and growth is a cure worse than the disease, as ecologists often argue, or the reverse, as stereotypical economists postulate in polarized debates.

While not denying that humans have treaded heavily on their life support systems, and acknowledging that growth cannot go on into the infinite future, Ausubel nonetheless combines the optimism of technologists about what is possible with that of economists about what has been accomplished. "Modern economies still work far from the limit of system efficiency," he argues, sounding very much like the technologists or ecologists who question the efficiency of markets. But this is good news, he goes on, since there is thus a great deal of room for dramatic "lightening of the load" as society catches up with what engineers have already invented. Rather than despairing from the evidence that societies implement only a small fraction of available efficient technologies and practices, Ausubel considers it our salvation. "The historical record reveals that the world has progressively lightened its energy diet over the past 200 years by favoring hydrogen atoms over carbon, in our hydrocarbon stew," he has written. "All these analyses imply that over the next hundred years, the human economy will clear most of the carbon from its system and move, via natural gas, to a hydrogen metabolism." Ausubel concludes, "We are heading in the right direction, toward micro-emissions. The way is long, but we are on the right path."[17]

What about land, the commodity so needed to grow food to feed the projected doubling or tripling of the population that most analysts expect, barring catastrophe, by the end of the next century? Here, too, Ausubel recognizes the threat to nature from the destruction of natural forests and wild lands, and does not deny the need to restore land for natural systems. But, citing a study led by the agronomist Paul Waggoner, Ausubel believes the solution again will be technological. Grain yields are so low in developing countries that if farmers in LDCs simply raise their productivity levels to half that in the United States or Europe, then "10 billion people on average can enjoy the diet 6,000 calories brings [comparable to wealthy diets today], and spare one quarter of the present 1.4 billion hectares of crop land. The quarter spared is about twice the size of Alaska [and about half the size of the Amazon basin]. If future farm land on average yielded today's U.S. corn, 10 billion eating an American diet could allow cropland the area of Australia to revert to wilderness."

But what of mountains of waste materials, toxic trash, and fuel dumps that launched the infamous odyssey of a garbage barge marooned on the high seas for weeks in 1988 in search of a willing landfill? Faced with this, Ausubel and colleagues Robert Frosch, Robert Ayres, and others, who coined the phrase "industrial ecology" to refer to the natural/social system in which humans are imbedded in nature, are less confident. One tenet of industrial ecology is "dematerialization" (the decline over time in the weight of materials used to meet a given economic function).[18] The rates of dematerialization would matter enormously for the environment, but are "equivocal," the industrial ecologists admit.

Let's contrast the overall optimism of these industrial ecologists with the views of some well-known natural ecologists (often labeled "doomsayers" by their opponents and in media debates): Paul and Anne Ehrlich of Stanford University and Norman Myers of Oxford, England. The Ehrlichs, for example, take long-term views and combine them with a biological twist to try to explain why so many people prefer optimism to pessimism, regardless of reality:

> People aren't scared because they evolve biologically and culturally to respond to short-term "fires" and to tune out long-term "trends" over which they have no control. Only if we do what doesn't come naturally—if we determinantly focus on what seems to be gradual or nearly imperceptible changes—can the outlines of our predicament be perceived clearly enough to be frightening.[19]

They are clearly frightened by rates of global change and spell out in painstaking detail a myriad collection of plausible disasters—famines, extinctions, epidemics—if we continue on present growth patterns.

At first blush, there seems to be the usual implacable division between economists and ecologists. Yet a deeper analysis of the economist/ecologist debate suggests that not all are as far apart as it might seem, although there are a number of paradigm differences. In fact, optimism or pessimism per se is not the important issue. Different analysts are focusing on the same set of issues, namely, that depending upon luck we may (1) end up with new technology leading to a stable and sustainable steady-state world one hundred years hence, or (2) with poor luck, we'll see spiraling catastrophic ecological and human crises because we ignore the externalities of our daily economy and the limits on indefinite, unsustainable expansion. The obvious difference is probability: stereotypical economists being optimistic that we'll use classical economic and political tools in time to become sustainable, ecologists being much more pessimistic about the resilience of nature and even more pessimistic about humans' capacity to visualize long-term ominous trends in time to reverse them successfully.

Ironically, perhaps, anyone who skims the myriad books written by Myers[20] and the Ehrlichs will find many virtually identical solutions to those suggested by optimistic technologists or economists: more efficient technologies, more rapid implementation, appropriate family planning activities, technology transfers from rich to poor, better educational systems, improved research and development on globally interconnected systems, and on and on with a very long list. Where is the fundamental debate? Is it simply one of style and taste about the likelihood of catastrophe? No, I think there really is some substantive disagreement on at least one fundamental point. Past performance and future potential of the economy are not enough if there is good reason to fear an ominous trend, the ecologists believe; there also needs to be concrete action to slow down the threats.

The gulf among ecological and economic paradigms primarily rests on two issues: (1) the likelihood that a vastly expanded scale of human numbers and technologies and productive activities can be undertaken in an ecologically sustainable way, and (2) even if ecosystem services were degraded, they're either relatively unimportant or could be substituted by other products of the human economy. Nearly all debaters who are responsible and honest agree that those questions are not definitely answerable, but they hold very different prior beliefs about the likelihood that the Earth will experience the optimistic or pessimistic outcomes. Technical or economic optimists admit that there are environmental risks and do suggest a host of policy actions to raise the rates at which human ingenuity can be brought to bear to help lower environmental risks. The same can be said for pessimistic technologists or ecologists, but they wish to see dramatic reversals in the overall scale of the human enterprise, in terms of both the number of people and the size of the economy, so as not to risk global crash from potentially unsustainable practices. While not all their solutions are identical, they hold a surprising number of them in common, even if obscured in a fog of optimistic versus pessimistic rhetoric.

PROCESS IS OUR MOST
IMPORTANT PRODUCT
....

It is a great error, I believe, to take either too much comfort or too much despair in believing that one side of the debate has a lock on the truth. The most likely outcome, I'll predict, is that both will turn out right in different specific cases: Some envisioned environmental crises will fade to insignificance, while others, as did the ozone hole, will spring out as nasty surprises that were barely perceived before they occurred. What is certain is that this pattern of good and bad luck will continue. I also believe that problems will become increasingly global and irreversible. It is urgent that we decide the extent to which we need to invest present resources to minimize potential future risks. Here is where the integrated assessment part of Earth systems science can play a useful role in the process of helping policy makers put decision making on a firmer, factual basis.

Let me reemphasize that I do not mean to imply that we take literally the quantitative conclusions of integrated assessment models that involve interconnections among many subcomponents, none of which can produce exact descriptions of the interconnected physical, biological, and social subsystems that must be integrated. Rather, the policy-making value of integrated assessment is *the process itself.* I would go so far as to suggest that it would be a dangerous delusion for the public or its representatives to make policy based solely on any analytic method, given the high degree of uncertainty in all the subcomponents that are interconnected in these society/nature models. However, the complexity that is represented by integrated assessment tools makes explicit and open to all those willing to learn what the logical consequences of those explicitly stated assumptions are for our environmental and economic future. A policy maker can learn a great deal about the possible behaviors of the real system by using integrated assessment tools as analytic toys, if you will, to help sharpen the intuitive grasp of the user about how complex interactions can amplify or reduce environmental or economic risks of a particular human activity or policy. The environmental benefits and economic costs of alternative tax policies can be analyzed, as can the environmental costs of dependence on coal rather than less carbon-intensive fuels.

In the end only a probability, usually subjectively decided, can be applied to any one particular outcome. But by engaging in the process, by playing the game of altering in the computer simulator the connections among various physical and biological parts of the system, or deliberately changing policies and asking the model to assess what the differential outcome of those policies would be, a policy maker becomes much better informed.

Nearly all policy making—be it medical, business, or governmental—is an intuitive value judgment about weighing risks and benefits of alternatives with imperfect information in most components. It is just as foolish to ignore the components of the problem that are amenable to quantification as it is to believe that simulation results addressing only those components are a rational, singular basis to choose the right policy. The learning is in the doing, not in "the answer." So, in this case, process is our most important product! The problem is finding policy makers willing to engage in such complex learning activities.

Decision making is an inherently social, not scientific, activity because it is a value-laden job. Even determining truth in science is a social activity to an extent most scientists don't like to admit, for in the short run at least, before the convincing experiments are performed, assessment of the state of the art of knowledge is done by tapping the wisdom of the community—or the prejudices of its elite, to phrase it in less laundered terms. This social activity guides what scientists do next, including what experiments are sanctioned—read funded—to be performed.

Who Is the Consensus?

Even though some social scientists and philosophers are fond of pointing out that, logically, science is just as irrational as any other activity where political power, opinions, and values creep in, I honestly believe that most scientific activities do strive consciously to minimize the decidedly unscientific fraternal behavior that characterizes, say, pork-barrel congressional actions. So practically, while not as value-free as its practitioners like to profess, science uses its objective tool, the scientific method, to test and retest its current prejudices—also known as theories or hypotheses. But when the state of science is marked by large uncertainties and the decision-making world is listening through ears called journalists, science becomes part of a grand (and sometimes not-so-grand) social activity.

Even if we agree to seek consensus, the problem of credibility remains: defining who is a member of the community whose wisdom should be tapped and whose opinions included as an entry in the collective description of the state of the art of science that then can be used for political purposes, like choosing the magnitude of a carbon tax or the increased public purchase of land for wildlife reserves and corridors between them as a hedge against rapid climate change. In other words, which experts get registered to vote in this scientific consensus-building election? There are plenty of opportunities here for polemics, hype, and character assassination by certain interests with a political or economic stake in the outcome. It will be up to us to learn how to see through this public relations blitz.

Those scientists insisting that we hold our tongues until the normal process of science reveals "truth"—a firm consensus estimate of what will happen—miss the fact that personal, corporate, and governmental decision making is almost always based on best-guess estimates for a range of outcomes, usually in the absence of a firm consensus. This is how most insurance gets purchased, investments made, and medical choices evaluated. While state-of-the-art assessment of expert opinion is an admittedly social function, and thus all of us who participate will, to some degree, be influenced by subjective opinion, so too is evaluation of military risk, medical risk, economic risks, future interest rates, and almost all the other socially important areas in which decisions need to be made in the face of large uncertainties.

The social problem of science policy making then becomes how to make nonexpert communities knowledgeable about what a broad cross section of relevant experts thinks might happen and what the probabilities of those occurrences are. Most important, of course, is some guidance on what the impacts of such occurrences could imply for environment and society. That consensus-assessment process is inherently a social activity—therefore not one for which the scientific method directly applies—and this makes many scientists edgy.

The scientific method, of course, is applied by individuals and groups trying to assess the validity of any particular theory or to estimate the probability of any particular outcome. But in the absence of unequivocal experiments, different individuals and groups are likely to have different expert opinions. At any snapshot in time, the evolving opinion of some knowledgeable community of experts is the scientific input needed by the social process of decision making. At some level it will be necessary to survey expert opinion, and to make science policy by using such consensus guesswork. Better to do it explicitly and formally with transparent methods than by special interests' press releases outlining the opinions of their favorite experts faxed to the halls of power all over the globe—what I often refer to in my lectures as the "one fax, one vote syndrome."

The Carnegie Mellon University's Engineering and Public Policy Department turned its frustration at the low level of sophistication in the public debate on global climate change into a scientific survey.[21] Sixteen scientists around the United States were asked in 1994 a series of questions designed to be as nonbiased as survey science permits. The scientists chosen were deemed by this team of integrated assessors (which had no clear prior persuasion as to whether the climate change problem was a mild or serious threat) as representatives of a knowledgeable cross section of experts with a wide range of views. One question, for example, asked us to draw a so-called cumulative probability distribution for the likelihood that, if carbon dioxide were to double, the world would eventually warm up by a given global averaged surface temperature. Fifteen of the sixteen scientists selected (and they included climatologists, oceanographers, and meteorologists, all who worked in the area) drew similar cumulative probability functions. While differing in quantitative detail, the fifteen graphs looked basically the same: all assigned a significant, but fairly small, probability of climate change that was either negative or less than 1°C. By small, I mean, 5 to 20 percent probability. In other words, fifteen of the sixteen scientists agreed that the contrarian view that climate change from human activities will be relatively small to negligible in the century ahead was a plausible outcome with a small probability. However, they assigned the bulk of the likelihood for global climate change in the next century to be somewhere between 1°C and 4°C—rather close to the consensus estimate in the standard national and international assessments that have been coming out for the past twenty years.

Rush to Denial

The community of scientists, that is fifteen of sixteen, all also agreed by their graphical presentation of cumulative probability distributions that there remained an unpleasant possibility, namely, that a nontrivial, but small, possibility existed that surprises could occur that could see climate changes greater than 4°C—some drawing their graphs out as far as 10°C warming or more in so-called surprise scenarios. Although they gave this possibility less than a 20 percent chance (as do I), a 10 percent chance of a potentially catastrophic event would certainly motivate most business leaders or individuals to buy insurance to exempt themselves from experiencing the full damages that such an outcome would entail or, better, to undertake actions to reduce the likelihood it would happen ("deterrence," in the parlance of the strategic security folks). But, of course, those value questions for policy makers were not asked of these scientists. Rather, they were simply asked to offer their technical expertise: assessing outcomes and best guesses of corresponding probabilities.

Who was the sixteenth scientist who didn't share the consensus? The repeated holdout was Richard Lindzen from MIT, the harsh critic of global-warming science and scientists.

Lindzen has derided the process of public discussion of the global-warming issue as unscientific and premature, echoed by the talk-show host Rush Limbaugh, editorials in the *Wall Street Journal*, *The Cato Business Review*, and slick videos (portraying CO_2 as good for the planet) produced by the coal lobby. Lindzen states that it is highly likely that there will be only negligible outcomes from global change. At first glance, a look at his graph showed the same shape as those of his fifteen peers, until one noticed that along the horizontal axis, the right-hand extreme number, rather than being 10°C as the maximum warming event, was 1°C. Lindzen said to the interviewers that in his scientific judgment, there was less than a 2 percent chance that warming from a doubling of CO_2 would be greater than 1°C. Recall now that the other fifteen scientists all agreed that a small warming (less than 1°C) was a possibility, and assigned odds of something like 5 to 20 percent typically to that outcome. So, none of the rest of us denied the wide range of uncertainty that includes negligible outcomes.

But Lindzen and a handful of other scientists, who seem to be given media time comparable to the bulk of the climatological community, repeatedly assert that somehow their special knowledge of the future allows them to know with high certainty something virtually everybody else in the expert community disputes: that the probability of any nonnegligible outcomes is virtually zero. I think that this complacent judgment is scientifically outrageous, in view of the wide range of uncertainties over feedback mechanisms that nobody can reliably pin down today. Moreover, it ignores the growing fingerprint evidence discussed earlier.

A Climate Copernicus?

It is always possible that the fifteen of us in the survey are wrong and Lindzen is right. In fact, we each gave that about a 10 percent or so chance ourselves. There is even some romance in the notion. Science fiction movies frequently portray a courageous scientist who stands up against the crowd and insists that some consensus-approved project would cause a crack in the Earth, create a radioactive monster, or lead to some other dread outcome contrary to conventional wisdom. And real-life stories often define heroes as those who stand up to conventional wisdom. Galileo and his telescope cracked both the theory of celestial spheres and the credibility of the church establishment that proclaimed it as truth. The power structure did not react passively either. Admiration is (eventually) expressed for those who changed the way we think about things that were dogma.

And well it should, for it is an act of courage to practice conscientious objection. After all, for how many centuries had Ptolemaic supporters with flowing robes and respectable positions in church and state upheld the geocentric view of the universe? Indeed, science by consensus can produce foolish or dangerously wrongheaded views. On that point, I agree with Dick Lindzen: It takes the open process of science, through its official doubting procedure known as the scientific method, to undo wrong ideas over time.

While it may be that for every Ptolemy there eventually comes a Copernicus, I'll also bet that for every true Copernicus, there are at least a thousand pretenders. And on most issues, conventional wisdom turns out to be more or less right. Unfortunately, the media and the political process too often give equal weight to nearly all conflicting credentialed Ph.D. claimants of truth, whereas the scientific assessment process—such as that practiced by the U.S. National Research Council, the UN Intergovernmental Panel on Climate Change, or, in the specific example given, the group at Carnegie Mellon—tries to quantify and isolate outlier opinions from the state-of-the-art mainstream. Once in a while, the outliers will be right. Their views must be heard, but not presented as equally likely.

Therefore, the best judgment that the social process of environment/development policy making can operate on is the collective judgment of a representative sample of the knowledgeable community on the range of outcomes and their scientific probabilities. And this process should be repeated fairly often, since new knowledge does come in rapidly and we should rethink our policy directions in light of such discoveries. I have referred to the latter as "the rolling reassessment process."

There may be a climate Copernicus out there who'll prove one day that the conventional wisdom is way off. But at the moment, I'll ask my politicians to bet with the majority opinion determined by sampling a broad cross section of knowledgeable experts. This survey process is decidedly not the media notion of balance, which frequently pits wildly opposing extremes against each other in implacable dispute, as if no one knowledgeable believed anything else to be possible and the relative likelihood of these outlier views were identical to all other possibilities. Nor is it the political balancing act that might see the experts chosen by elliptical political opponents—say, Ralph Nader and Newt Gingrich in the United States. It is impossible to have a rational public debate about appropriate policy actions in such a state of conflict, frenzy, and distortion.

WHAT ARE SOME ACTIONS TO CONSIDER?

Decision making, I have repeated often, is a value judgment about how to take risks—gambling, if you will—in the environment-development area. Despite the often bewildering complexity, making value choices does not require a Ph.D. in statistics, political science, or geography. Rather, citizens need simple explanations using common metaphors and everyday language that ordinary people can understand about the terms of the debate. Once the citizens of this planet become aware of the various trade-offs involved in trying to choose between business-as-usual activities and sustainable environmental stewardship, the better will be the chances that the risk-averse common sense of the "average" person may be thrust upon reluctant politicians and power brokers by a public that cares about its future and that of its planet, and knows enough not to be fooled by simple solutions packaged in slick commercials or editorials by any special interest.

What are the kinds of actions that can be considered to deal with global change problems like climate change? The following list is a consensus from a multidisciplinary business, university, and government assessment conducted by the National Research Council in 1991. It is encouraging that this ideologically diverse group (including the economist Nordhaus, the industrialist Frosch, and the climatologist Schneider) could agree that the United States, for example, could reduce or offset its greenhouse-gas emissions by between 10 and 40 percent of 1990 levels at low cost, or at some net savings, if proper policies were implemented. Here is the Council's suggested list:

1. Continue the aggressive phaseout of CFC and other halocarbon emissions and the development of substitutes that minimize or eliminate greenhouse gas emissions.
2. Study in detail the "full social cost pricing" of energy, with a goal of gradually introducing such a system. On the basis of the principle that the polluter should pay, pricing of energy production and use should reflect the full costs of the associated environmental problems.
3. Reduce the emissions of greenhouse gases during energy use and consumption by enhancing conservation and efficiency.
4. Make greenhouse warming a key factor in planning for our future energy supply mix. The United States should adopt a systems approach that considers the interactions among supply, conversion, end use, and external effects in improving the economics and performance of the overall energy system.
5. Reduce global deforestation.
6. Explore a moderate domestic reforestation program and support international reforestation efforts.
7. Maintain basic, applied, and experimental agricultural research to help farmers and commerce adapt to climate change and thus ensure ample food.
8. Make water supply more robust by coping with present variability by increasing efficiency of use through water markets and by better management of present systems of supply.

9. Plan margins of safety for long-lived structures to take into consideration possible climate change.
10. Move to slow present losses in biodiversity.
11. Undertake research and development projects to improve our understanding of both the potential of geo-engineering options to offset global warming and their possible side effects. This is not a recommendation that geoengineering options be undertaken at this time, but rather that we learn more about their likely advantages and disadvantages.
12. Control of population growth has the potential to make a major contribution to raising living standards and to easing environmental problems like greenhouse warming. The United States should resume full participation in international programs to slow population growth and should contribute its share to their financial and other support.
13. The United States should participate fully with officials at an appropriate level in international agreements and in programs to address greenhouse warming, including diplomatic conventions and research and development efforts.

This is a remarkable list, considering the diversity of the participants' backgrounds and their varying ideological perspectives. But in the crucible of open debate in front of an expert community, self-interest polemics and media grandstanding are ineffective, if not counterproductive. This group didn't assert that catastrophe was inevitable, nor that it was improbable. We simply believed that prudence dictates that "despite the great uncertainties, greenhouse warming is a potential threat sufficient to justify action now." Integrated assessments of the policy options offered by the National Research Council Report are actively being pursued with a variety of models.

But even this comprehensive list of thirteen recommendations still ignored two fundamental aspects: the desperate need for (1) an intelligent, nonpolemical public debate about global change, and (2) interdisciplinary public education that also teaches students about whole systems and long-term risk management, not only traditional areas of isolated specialization. Without these elements it will be difficult to get the public solidly behind policies that invest immediate national resources for long-term global goals whose necessity is not well understood.

Environment and (or versus) Development

While the NRC report did acknowledge the importance of international dimensions of global change policy making, it was still largely a developed-country perspective. Developing countries often have very different perspectives than LDCs. First of all, LDCs are struggling to raise literacy rates, lower death rates, increase life expectancy, provide employment for burgeoning populations, and reduce local air and water pollution that pose imminent health hazards to their citizens and environments. Protecting species or slowing climate change are simply low on their priority lists as compared to more mature economic powers. It is ironic, if understandable, that LDCs put abatement of global change disturbances low on their priority lists, despite the fact that nearly all impact assessments suggest that it is these very countries that are most vulnerable to climatic change.

There is a phrase in economics known as the "marginal dollar." In our context it means that given all the complexity of interconnected physical, biological, and social systems, what is the best place to invest the next available dollar so as to bring the maximum social benefit? I have heard many representatives of LDCs exclaim that until poverty is corrected, preventable disease stamped out, injustice redressed, and economic equity achieved, they will invest their precious resources on these priorities. My response has been that climatic changes can exacerbate all of those problems they rightly wish to address, and thus we should seek to make investments that both reduce the risks of climate change and help with economic development (transfer of efficient technologies being a prime example). It is a great mistake to get trapped in the false logic of the mythical "marginal dollar," for it is not necessary that every penny of the next available dollar go exclusively to the highest-priority problem, whereas all the rest must wait until Priority 1 is fully achieved. To me, the first step is to get that marginal dollar cashed into small change, so that many interlinked priority problems can all be at least partially addressed. Given the large state of uncertainty surrounding the costs and benefits of many human and natural events, it seems most prudent to address many issues simultaneously and to constantly reassess which investments are working and which problems—including global change—are growing more or less serious (the rolling reassessment process).

It takes resources to invest, of course, and since the bulk of available capital is in developed countries, it will require international negotiations—"planetary bargaining," it has been called—to balance issues of economic parity and social justice with environmental protection. Such negotiations are under way under UN auspices, and will likely take many years to work out protocols that weigh the diverse interests and perceptions of the world's nations.[22]

While most of the National Research Council group's recommendations just cited are directed at the level of national governments in developed countries, a parallel set applies to individuals and smaller-scale institutions. I know it sounds preachy—it's like telling people that their vote could make a difference, even though everyone knows nobody wins elections by one vote—but turning the lights or TV or computer off when you leave the room is important. For example, a number of elections would have been different if only one more person in each precinct had voted for the loser. Likewise, when five billion people conserve energy this way, a thousand times each in a year, the savings add up. Moreover, doing so sets the right tone for a lot of the small lifestyle changes we need to make if we are going to be effective in slowing global change.

For example, if you want to buy a refrigerator and you find two models that look identical with similar features, but one costs $1,000 and the other $900, which one are you likely to buy? The cheaper one, probably. But if you read the label, you might find that the $1,000 refrigerator uses slightly less energy than the $900 refrigerator. Probably it costs more money because it's better insulated. All we need to do is a little mental arithmetic to calculate from the information given on the label just how much we might save on our electric bill each year from the better-insulated refrigerator. Let's say it's twenty-five dollars a year. That means in four years, we've gotten our hundred dollars back, and for the ten- to fifteen-year lifetime that is typical of refrigerators, we will have actually come out ahead. At the same time, we will have helped the environment, since using less energy means less pollution. Currently, this nonmarket amenity is not part of the product price, even if it is a real social cost.

We all need to replace our cars. The next time we do that, we can read the labels for fuel efficiency. Is it really so important to have the biggest and the fastest car? Why not do something that's good for the environment and our pocketbooks at the same time? Get a more energy-efficient car.

Politicians are remarkably good at reacting to the perceptions of their constituents—when we communicate our feelings. If we are going to influence political leaders to be creative and work for long-term solutions to help us use energy efficiently at home, to provide alternatives to the Chinese to advance past their planned inefficient use of dirty coal, and to help the Indonesians and Brazilians reverse their rapid deforestation, the politicians have got to know we want them to. Let them know your concerns and values. If we are silent, only the faxes from the special-interests get through.

That is, in fact, the bottom line of this book: It is the job of every citizen to choose how to balance the environmental risks of certain kinds of economic development against the perceived benefits, and to become well enough informed to be able to make the value judgments implied in that balancing act. Experts across the many disciplines that comprise Earth systems science can help say what might happen and at what probability. The next time you hear experts telling you "what to do," remember to ask them these three questions: (1) What can happen? (2) What are the odds? and (3) How do you know? Make sure you get them to separate out what parts of their expert judgments deal with well-established, objective probabilities and what parts are subjective. Expertise ends there. What to do is the responsibility of all of us, if only we'd take it.

And what if we don't act responsibly? In that case we're gambling that what it would cost us to protect the planet will, by some good luck, turn out to be more costly than the damages we carelessly inflicted on the Earth. This is a gamble that we—and the other living creatures who share our planet, but not in the decision making—simply can't afford to lose.

NOTES

....

INTRODUCTION: IT'S A MATTER OF SCALE
....

1. S. A. Levin. 1992. The problem of pattern and scale in ecology. *Ecology* 73: 1943–67.
2. T. L. Root and S. H. Schneider. 1995. Ecology and climate: Research strategies and implications. *Science* 269: 334–41. These authors argue that global change problems are best addressed with multiscale, multidisciplinary, and multi-institutional approaches.
3. W. G. Ernst, ed. In press. *Earth Systems.* New York: Cambridge University Press. A textbook example of the breadth of multidisciplinary knowledge needed to address earth systems issues.
4. R. Peters and T. Lovejoy, eds. 1992. *Global Warming and Biological Diversity.* New Haven, Conn.: Yale University Press.
5. P. R. Ehrlich and J. P. Holdren. 1971. Impact of population growth. *Science* 171: 1212–17.
6. R. Cantor and S. Rayner. 1994. Changing perceptions of vulnerability. In *Industrial Ecology and Global Change.* R. Socolow, C. Andrews, F. Berkhout, and V. Thomas, eds. Cambridge: Cambridge University Press, pp. 69–83.
7. National Academy of Sciences. 1991. *Policy Implications of Greenhouse Warming.* Washington, D.C.: National Academy Press.

CHAPTER I. THE ORGANIC AND NONLIVING EARTH: A DYNAMIC COHESION
••

1. C. Sagan and G. Mullen. 1972. Earth and Mars: Evolution of atmospheres and temperatures. *Science* 177: 52–56.

2. W. Broecker. 1990. *How to Build a Habitable Planet*. Palisades: Lamont-Doherty Geological Observatory Press. A good source for learning about the geochemical fundamentals of our planet, written by one of the most insightful earth scientists of our time.

3. J. F. Kasting. 1993. Earth's early atmosphere. *Science* 259: 920–26. Contains citations and perspective on earlier works.

4. J. E. Lovelock. 1995. *The Ages of Gaia: A Biography of Our Living Earth*. New York: Norton. Jim Lovelock's most recent update and viewpoint.

5. S. H. Schneider and P. Boston, eds. 1991. *Scientists on Gaia*. Cambridge, Mass.: MIT Press. The edited and refereed proceedings of the first "establishment" scientific conference on the Gaia hypothesis. That seminal event is recounted in most subsequent popular books on Gaia as a springboard for serious scientific debate. For example, see L. E. Joseph. 1990. *Gaia: The Growth of an Idea*. New York: St. Martin's.

6. J. E. Lovelock and L. Margulis. 1973. Atmospheric homeostasis by and for the biosphere: The Gaia hypothesis. *Tellus* 26: 2, is the classical scientific article on Gaia.

7. D. Schwartzman, M. McMenamin, and T. Volk. 1993. Did surface temperature constrain microbial evolution? *Bioscience* 43: 390–93.

8. L. A. Frakes. 1979. *Climates through Geologic Time*. Amsterdam: Elsevier.

9. C. J. Allegre and S. H. Schneider. 1994. The evolution of the Earth. *Scientific American* 241: 44–51.

10. J. Imbrie and K. P. Imbrie. 1979. *Ice Ages: Solving the Mystery*. Short Hills, N.J.: Enslow. A wonderful source of science and history dealing with the ice age puzzle.

11. E. O. Wilson. 1992. *The Diversity of Life*. New York: Norton. Another informative and accessible science popularization by this Pulitzer Prize-winning ecologist.

CHAPTER 2. THE COEVOLUTION OF
CLIMATE AND LIFE
••••••••••••••••••••••

1. W. H. Schlesinger. 1991. *Biogeochemistry: An Analysis of Global Change.* New York: Academic Press.

2. S. H. Schneider. 1994. Detecting climatic change signals: Are there any "fingerprints"? *Science* 263: 341–347. Reviews the history of the aerosol-climate debate and offers many additional references. This paper laid out the reasoning that allowed subsequent assessments to express increased confidence in the detection of global warming effects in the climate record.

3. Intergovernmental Panel on Climatic Change (IPCC), 1996. *Climate Change 1995. The Science of Climate Change: Contribution of Working Group I to the Second Assessment Report of the Intergovernmental Panel on Climate Change.* J. T. Houghton, L. G. Meira Filho, B. A. Callander, N. Harris, A. Kattenberg, and K. Maskell, eds. Cambridge: Cambridge University Press. See chapter 10 for a review of ocean carbon chemistry. Hereafter cited as IPCC 1996, WG I.

4. E. J. Barron, P. J. Fawcett, D. Pollard, and S. L. Thompson. 1993. Model simulations of Cretaceous climates: The role of geography and carbon dioxide. *Philosophical Transactions of the Royal Society of London* 341: 307–16.

5. L. F. Richardson. 1922. *Weather Prediction by Numerical Processes.* Cambridge: Cambridge University Press. The classical and visionary precursor to the weather and climate models.

6. Richardson. *Weather Prediction by Numerical Processes*, pp. 219–20.

7. P. N. Edwards. 1996. *The Closed World: Computers and the Politics of Discourse in Cold War America.* Cambridge, Mass.: MIT Press.

8. S. H. Schneider and R. Londer. 1984. *The Coevolution of Climate and Life.* San Francisco: Sierra Club. Chapter 6 provides a layperson's overview of climate modeling.

9. S. H. Schneider, S. L. Thompson, and E. J. Barron. 1985. Mid-Cretaceous, continental surface temperatures: Are high CO_2 concentrations needed to simulate above-freezing winter conditions? In *The Carbon Cycle and Atmospheric CO_2: Natural Variations Archaen to Present.* E. Sundquist and W. Broecker, eds. Geophysical Monograph Series, Vol. 32, American Geophysical Union, Washington, D.C., pp. 554–59.

10. R. A. Berner, A. C. Lasaga, and R. M. Garrels. 1988. The carbonate-silicate geochemical cycle and its effect on atmospheric carbon dioxide over the past 100 million years. *American Journal of Science* 283: 641–83.

11. M. I. Budyko, A. B. Ronov, and A. L. Yanshin. 1987. *History of the Earth's Atmosphere.* New York: Springer-Verlag. Although their conclusions are controversial, they offer what they believe to be direct evidence for high CO_2 100 million years ago.

12. For sources of the data on ice sheets, see Schneider and Londer, *The Coevolution of Climate and Life,* chapter 3.

13. J. A. Eddy and H. Oeschger, eds. 1993. *Global Changes in the Perspective of the Past.* New York: Wiley. Articles on many aspects of the ice age story, including the problem of trying (mostly in vain) to find paleoclimatic analogs to future anthropogenic climatic changes. Another excellent source for discussion of ice age theories is the text, T. J. Crowley and G. R. North. 1991. *Paleoclimatology.* New York: Oxford University Press.

14. C. Lorius, J. Jouzel, D. Raynaud, J. Hansen, and H. Le Treut. 1990. The ice-core record: Climate sensistivity and future greenhouse warming. *Nature* 347: 139–45. A discussion of these data and their possible bearing on the modern global warming debate.

15. R. A. Berner. 1993. Paleozoic atmospheric CO2: Importance of solar radiation and planet evolution. *Science* 261: 68–70.

16. M. I. Hoffert and C. Covey. 1992. Deriving global climate sensitivity from paleoclimate reconstructions. *Nature* 360: 573–76.

17. For citations and a fuller discussion, see Schneider and Londer, *The Coevolution of Climate and Life*, p. 233.

CHAPTER 3. WHAT CAUSES CLIMATE CHANGE?
....

1. The data for the trend in twentieth-century surface temperature come from IPCC 1996, WG I, an international assessment by 100 scientists. Chapter 3 presents a large array of recent evidence for climate change, including a discussion of how the much-exaggerated apparent differences (e.g., S. F. Singer. 1995. Letter to *New York Times,* "Global Warming Remains Unproved," Sept. 19, 1995) between the

17 years of satellite-derived mid- to upper atmospheric temperatures and directly measured surface temperatures can be reconciled (e.g., see pp. 147–48 of IPCC). Passionate debate about the validity of these records is ongoing. E.g. see the article by C. Prabhakara, J.-M. Yoo, S. P. Maloney, J. J. Nucciarone, A. Arking, M. Cadeddu, and G. Dalu. 1996. "Examination of 'Global Atmospheric Temperature Monitoring with Satellite Microwave Measurements': (2) Analysis of Satellite Data," *Climatic Change* 33: 459–76; a critical comment on it by the authors of the satellite technique, R. W. Spencer, J. R. Christy, and N. C. Grody. 1996. "Analysis of 'Examination of "Global Atmospheric Temperature Monitoring with Satellite Microwave Measurements,"'" *Climatic Change* 33: 477–89; and Prabhakara and colleagues' defense. 1996. "Examination of 'Global Atmospheric Temperature Monitoring with Satellite Microwave Measurements': (3) Cloud and Rain Contamination," *Climatic Change* 33: 491–96; the bottom line of which is that the satellite technique still has not satisfactorily accounted for distorting effects of tall rain clouds over extensive parts of tropical oceans and thus has not yet been proved to provide a fully calibrated temperature trend record for a known segment of the atmosphere. Thus, it is my opinion that the surface temperature thermometer network, because it is both a long-term record (some ten times longer than the satellite data) and a measure of climate where it is most important for humans and nature (at the surface), is still the best measure yet available for climatic inferences. The two techniques complement each other, and resolutions to the disputes cited are ongoing.

2. See chapters 2 and 3 of IPCC 1996, WG I; and S. H. Schneider. 1994. Detecting climatic change signals: Are there any "fingerprints"? *Science* 263: 341–47.

3. K. E. Trenberth and J. W. Hurrel. 1996. The 1990–1995 El Niño-Southern Oscillation event: Longest on record. *Geophysical Research Letters* 23: 57.

4. E. N. Lorenz. 1968. Climatic determinism in causes of climatic change. *Meteorological Monographs* 8: 1–3.

5. J. Gleick. 1987. *Chaos.* New York: Viking Press.

6. J. A. Hansen, R. Ruedy, and M. Sato. 1992. Potential climate impact of Mt. Pinatubo eruption. *Geophysical Research Letters* 19: 215–18. Using the same climate model tapped for global warming forecasts, these authors pre-

dicted a surface cooling temperature of a few tenths of a degree Celsius. Chapter 3 of IPCC 1996, WG I confirms that the prediction was realistic.

7. K. E. Trenberth, ed. 1992. *Climate System Modeling*. Cambridge: Cambridge University Press.

8. One example of confusing short-term limits to the predictability of weather with forecasts of long-term changes in climate statistics is: M. L. Parsons. 1995. *Global Warming: The Truth Behind the Myth*. New York: Plenum. As is often the case in "truth vs. myth" books, zealous authors get truth and myth reversed, as I noted in a book review: S. H. Schneider. 1996. Climate reversal. *Nature* 381: 384–86. For an exposé of a number of prime "contrarians" and their supporters, see R. Gelbspan. Dec. 1995. The heat is on. *Harpers* 35. Gelbspan is a Pulitzer Prize-winning journalist.

9. S. H. Schneider. 1993. Degrees of certainty. *National Geographic Research and Exploration* 9(2): 173–90.

10. W. M. Washington and C. L. Parkinson. 1986. *An Introduction to Three-Dimensional Climate Modeling*. New York: Oxford University Press.

CHAPTER 4. MODELING HUMAN-INDUCED GLOBAL CLIMATE CHANGE
....

1. A. Franzén. 1992. *Vasa: The Brief Story of a Swedish Warship from 1628*. Stockholm: Bonniers and Norstedt.

2. For fuller discussion, see IPCC 1996, WG I chapter 5

3. Critics of climate model parameterizations, such as Richard Lindzen of MIT, often argue that since physical processes known to be important at one scale (that of a thunderstorm, Lindzen likes to argue) and are not explicitly included in GCMs with large grid boxes, the GCMs must thus, necessarily he argues, be invalid (e.g., R. S. Lindzen. 1990. Some coolness to global warming. *Bulletin of the American Meteorological Society* 71[3]: 288–99). I have commented that it is never possible, nor likely to be necessary, to include all small-scale processes, only to test to see which small-scale processes significantly contribute to large-scale effects (e.g., see T. L. Root and S. H. Schneider. 1995. Ecology and climate: Research strategies and implications. *Science* 269: 334–41). I also have cited data suggesting that models, even if by luck, were roughly cor-

rectly predicting large-scale upper atmospheric humidity increases when surface temperatures increased (S. H. Schneider. 1991. Response to Hugh Ellsaesser. *Bulletin American Meteorological Society* 72: 1009–11); this suggests that Lindzen's small-scale analysis was not likely to be very significant for large-scale predictions. Think of a thunderstorm like a mushroom—intense stems of high humidity, updrafts and precipitation, surrounded by a cap of dry sinking air. Accounting for such dry air, Lindzen contends, will reduce the sensitivity of the climate by up to a factor of ten relative to GCM predictions. My counter-argument is backed up by the large-scale data I cited above and also other data mentioned by D. Rind, E.-W. Chiou, W. Chu, J. Larsen, S. Oltmans, J. Lerner, M. P. McCormick, and L. McMaster. 1991. Positive water vapour feedback in climate models confirmed by satellite data. *Nature* 349: 500–503. I contend that one cannot simply apply theory and observations at the scale of one storm to the collective effects of the interactions of many storms. Each "mushroom" is not an entity unto itself; rather, the downdrafts of one "cap" interact with the motions of its neighbors, etc. Thus, the large-scale relationships between surface temperature and the humidity profile in the atmosphere may be very different from what one would infer by studying one thunderstorm in isolation. Root and I argued that only empirical tests *at the scale of GCM grids* can help to validate the performance of GCMs—not assertions that known small-scale processes are not explicitly included. This kind of debate is very confusing to lay audiences, but scientific assessments by many experts, like the IPCC, are well aware of these problems and conclude, nonetheless, that for a variety of reasons, GCM estimates of global climate sensitivity are likely to be valid to a factor of 2 or 3. Thus the warming range is typically given as 1.5 to 4.5° if CO_2 were to double.

4. H. E. Wright, J. E. Kutzbach, T. Webb III, W. F. Ruddiman, F. A. Street-Perrott, and P. J. Bartlein, eds. 1993. *Global Climates Since the Last Glacial Maximum*. Minneapolis: University of Minnesota Press. Contains a wealth of information about both paleoclimatic and paleoecological records and their relationships to possible causal factors such as orbital variations or changes in CO_2 concentration over geological time.

5. Wright et al., *Global Climates Since the Last Glacial Maximum*.

6. A. Berger. 1992. *Le Climat de la terre: un passé pour quel avenir?* Brussels: De Boeck Université.

7. Wright et al., *Global Climates Since the Last Glacial Maximum*.

8. See chapter 2 of IPCC 1996, WG I, especially p. 117, and the debate among several authors: Testing for bias in the climate record. *Science* 271: 1879–83.

9. S. H. Schneider. 1994. Detecting climatic change signals: Are there any "fingerprints"? *Science* 263: 341–347.

10. See chapter 8 of IPCC 1996, WG I for a very thorough discussion of climatic signal detection and attribution issues. This chapter and its lead author, Ben Santer, became the focus of a series of character attacks from an industry lobby, The Global Climate Coalition, and a few veteran "contrarian" scientists, leading to a spate of media stories alleging "scientific cleansing" on the one hand, and deliberate obfuscation on the other. See, for example E. Masood. 1996. Climate Report "Subject to scientific cleansing." *Nature* 381: 546; or W. K. Stevens. 1996. At hot center of debate on global warming. *New York Times*, August 6, 1996, pp. B5–B6.

11. B. D. Santer, K. E. Taylor, T. M. L. Wigley, T. C. Johns, P. D. Jones, D. J. Karoly, J. F. B. Mitchell, A. H. Oort, J. E. Penner, V. Ramaswamy, M. D. Schwarzkopf, R. J. Stouffer, and S. Tett. 1996. A search for human influences on the thermal structure of the atmosphere. *Nature* 382: 39–46.

12. IPCC 1996, WG I, p. 5.

13. See S. H. Schneider and Lynne Mesirow. 1976. *The Genesis Strategy: Climate and Global Survival*, New York: Plenum. I can report with some pride that in this book I debated the relative effects of natural fluctuations with human-produced aerosol cooling and greenhouse gas warming, noting on page 11 that "climatic theory is still too primitive to prove with much certainty whether the relatively small increases in CO_2 and aerosols up to 1975 were responsible for this climate change. I do believe, however, that if concentrations of CO_2, and perhaps of aerosols, continue to increase, demonstrable climatic changes could occur by the end of this century, if not sooner; recent calculations suggest that if present trends continue, a threshold may soon be reached after which the effects will be

unambiguously detectable on a global basis. Problemati-
cally, by that point it may be too late to avoid the dangerous consequences of such an occurrence, for *certain proof*
of present theories can come only *after* the atmosphere
itself has 'performed the experiment.'" Writing twenty-one
years later, I think I would change a couple of words, like
"unambiguous," since the signal detection problem is a
matter of probability and some analysts are satisfied that
enough coincidences already exist to suggest "detection,"
whereas others want to reduce the possibility of chance
still more. The threshold of probability for detection is thus
a judgment, not an objective determination.

14. Chapter 8 of IPCC 1996, WG I summarizes the situation well.
15. S. Manabe and R. J. Stouffer. 1993. Century scale effects of
    increased atmospheric CO_2 on the ocean-atmosphere sys-
    tem. *Nature* 364: 215–18.
16. S. L. Thompson and S. H. Schneider. 1982. CO_2 and cli-
    mate: The importance of realistic geography in estimating
    transient response. *Science* 217: 1031–33.
17. This last paragraph of the IPCC 1996, WG I, Executive Sum-
    mary, p. 7, was not controversial; it was adopted by the ple-
    nary group with little dissent and minimal rephrasing of the
    original draft text. I was especially gratified for two reasons:
    (1) most other paragraphs of the Executive Summary were
    debated at tedious length, and (2) I was responsible for
    drafting the original text presented to the full plenary.
18. W. S. Broecker. 1994. Massive iceberg discharges as trig-
    gers for global climate change. *Nature* 372: 421–24; J.
    Imbrie and K. P. Imbrie. 1979. *Ice Ages: Solving the Mys-
    tery*. Short Hills, N.J.: Enslow.
19. Not surprisingly, IPCC 1996, WG I provides a good brief
    summary of the flip-flop controversy, too; see pp. 177–79.
20. S. H. Schneider. 1995. The future of climate: Potential for
    interaction and surprises. In *Climate Change and World
    Food Security*. T. E. Downing, ed. NATO ASI Series: Ser. I,
    Global Environmental Change, vol. 37. Heidelberg: Springer-
    Verlag, pp. 77–113.

CHAPTER 5. BIODIVERSITY AND
THE BATTLE OF THE BIRDS

1. C. Darwin. 1859. *On the Origin of Species by Means of Natural Selection*. London: John Murray.

2. M. B. Davis. 1976. Pleistocene biogeography of temperate deciduous forests. *Geoscience and Man* 13: 13–26.

3. J. T. Overpeck, R. S. Webb, and T. Webb III. 1992. Mapping eastern North American vegetation change over the past 18,000 years: No analogs and the future. *Geology* 20: 1071–74.

4. R. W. Graham and E. C. Grimm. 1990. Effects of global climate change on the patterns of terrestrial biological communities. *Trends in Ecology and Evolution* 5(2): 89–92. See also Graham's readable entry on Animals and Climate, pp. 27–32, in S. H. Schneider, ed. 1996. *Encyclopedia of Climate and Weather, A-K.* New York: Oxford University Press.

5. P. R. Ehrlich and J. Roughgarden. 1987. *The Science of Ecology.* New York: Macmillan. A useful reference on ecology. See also S. L. Pimm. 1991. *The Balance of Nature: Ecological Issues in the Conservation of Species and Communities.* Chicago: University of Chicago Press.

6. T. L. Root. 1988. Environmental factors associated with avian distributional boundaries. *Journal of Biogeography* 15: 489–505.

7. T. L. Root and S. H. Schneider. 1995. Ecology and climate: Research strategies and implications. *Science* 269: 334–41. The authors cite many studies that debate the effects of altered photosynthesis from extra CO_2 in the air.

8. E. O. Wilson. 1992. *The Diversity of Life.* New York: Norton.

9. R. M. May. 1994. Past efforts and future prospects towards understanding how many species there are. In *Biodiversity and Global Change.* O. T. Solbrig, H. M. van Emden and P. G. W. J. van Oordt, eds. Wallingford, Conn.: CAB International, pp. 71–84.

10. An editorial by Simon and the late U. C. Berkeley political scientist A. Wildavsky sported the provocative title "Facts, not species, are periled" and appeared on p. A23 of the May 23, 1993, *New York Times.* Simon and Wildavsky's strident insistence that ecological data refute island biography theory based predictions of bird extinctions was echoed by writer Stephen Budiansky in a 1994 letter in *Nature* 370: 105.

11. S. L. Pimm and R. A. Askins. 1995. Forest losses predict bird extinctions in eastern North America. *Proceedings of the National Academy of Sciences* 92(9): 343–47.

12. P. Vitousek. 1994. Beyond global warming: Ecology and global change. *Ecology* 75: 1861–76. This article also points out the risks of synergistic disturbances.

13. N. Myers and J. Simon. 1994. *Scarcity or Abundance: A Debate on the Environment.* New York: Norton, offers numerous examples of the radically different world views of data-oriented economists versus theory-oriented ecologists.

14. Strictly speaking, the avoidance of low probability but high consequence outcomes (i.e., "risk aversion") is not logically inconsistent with optimization of economic efficiency if one puts a high value on minimizing large risks. However, this is not the usual practice in cost/benefit calculations, which stress best-guess outcomes rather than extreme events.

CHAPTER 6. INTEGRATED ASSESSMENTS OF POLICY OPTIONS
··························

1. The best available impact assessment has been done by the hundreds of experts gathered under the auspices of the U.N.-sponsored Intergovernment panel on Climate Change, Working Group II (Working Group I is the climate effects group repeatedly cited in this book as IPCC 1996, WG I): Intergovernmental Panel on Climatic Change (IPCC), 1996. *Climate Change 1995. Impacts, Adaptations and Mitigation of Climate Change: Scientific-Technical Analyses. Contribution of Working Group II to the Second Assessment Report of the Intergovernmental Panel on Climate Change.* R. T. Watson, M. C. Zinyowera, and R. H. Moss, eds. Cambridge: Cambridge University Press. Even though this may be the best available assessment of climatic impacts, there are several serious omissions. For one, the animals focused on were livestock, fish in fisheries, and humans (consequences to their health from heat stress or disease vectors changed by climate). Almost entirely excluded from careful analysis was half of the biological kingdom: insects and wildlife. Ecosystems were explicitly treated, but primarily as vegetation biomes, rather than plant species. Issues like the tearing apart of community structure or ecological services (as I discussed in chapter 5) were given little attention. Such non-market impacts are difficult to assess quantitatively, as we saw in chapter 5, but need to be considered much more in future assessments. One compact review of

this problem of valuing "non-market" goods is R. V. Ayres. 1992. Assessing regional damage costs from global warming. In *The Regions and Global Warming: Impacts and Response Strategies.* J. Schmandt and J. Clarkson, eds. New York: Oxford University Press, pp. 182–98.

2. National Academy of Sciences. 1991. *Policy Implications of Greenhouse Warming.* Washington, D.C.: National Academy Press.

3. Ecological economists now have their own society and their own journal, *Ecological Economics.* The basic premise is that the economy is not apart from nature and degradation of natural capital (e.g., soil stocks) should be included in measures of welfare, not just standard accounting items like gross national product, consumption, and savings. They are especially concerned about the magnitude or scale of the human appropriation of such natural capital. One good entry point into the field is a collection of essays: H. E. Daly and K. N. Townsend, eds. 1993. *Valuing the Earth.* Cambridge, Mass.: MIT Press.

4. J. P. Holdren. 1991. Population and the energy problem. *Population and Environment* 12: 231–55, and Bongaarts, J. 1992. Population growth and global warming. *Population and Development Review* 18: 299–319. Both offer alternative scenarios of population, energy, technology and affluence for the twenty-first century.

5. Our initial calculations showed that under favorable assumptions, lowering the price of alternative or "backstop" energy systems could actually save more money in the long term than would be lost from the extra costs of fossil fuels if carbon taxes were imposed to "internalize" the suspected external damage to the climate. Unfortunately, the political reality is that the costs would fall to the present generation and the benefits to future generations—not an ideal situation for sitting politicians thinking of reelection. We also realized that our favorable assumptions are not the only plausible set, and that we need to account for economic losses that would be felt as capital currently invested in conventional energy enterprises is withdrawn and reinvested in efficiency or backstop technologies because a carbon tax raised the price of fossil fuels, making that class of fuels less economically attractive. On balance, I believe there would be a long-term bonus from climate policies like a carbon tax that are not currently included in most integrated assessment tools, but the

amount of that bonus is currently highly uncertain. Meanwhile, conventional analyses prevail, such as: D. Gaskins and J. Weyant. 1993. EMF–12: Model comparisons of the costs of reducing CO_2 emissions. *American Economic Review* 83: 318–23.

6. A special issue of the journal *Climatic Change*, to be published in late 1996, contains several articles and critical opinions on the uses and abuses of integrated assessment that can provide interested readers with more in-depth discussions and citations to the primary literature.

7. IPCC, 1996, WG I, chapter 7, discusses the sea level rise issue in depth.

8. J. G. Titus and V. Narayanan. 1996. The risk of sea level rise. *Climatic Change* 33: 151–212. These authors made an analysis in which the opinions of various experts on uncertain parameters in the models were aggregated as if each expert was equally qualified. This assumption is criticized in an editorial in the same issue of *Climatic Change* (145–49) by decision analyst Elizabeth Paté-Cornell, and the collection of papers in this issue are a good entry point to the literature.

9. M. G. Morgan and H. Dowlatabadi. In press. Learning from integrated assessment of climate change. *Climatic Change*.

10. J. Rotmans. 1994. *Global Change and Sustainable Development: A Modelling Perspective for the Next Decade.* National Institute of Public Health and Environmental Protection (RIVM), RIVM-report no. 461502004, Globo Report Series, Bilthoven, Netherlands. Also: J. Alcamo, ed. 1994. *Image 2.0: Integrated Modeling of Global Climate Change.* Dordrecht: Kluwer Academic.

11. W. D. Nordhaus. 1992. An optimal transition path for controlling greenhouse gases. *Science* 258: 1315–19; see also critical Letters by S. H. Schneider, H. Dowlatababi, L. Lave, and M. Oppenheimer, in *Science* (1993) 259: 1381–84, with a response by Nordhaus.

12. S. H. Schneider. 1995. The future of climate: Potential for interaction and surprises. In *Climate Change and World Food Security.* T. E. Downing, ed. NATO ASI Series: Ser. I, Global Environmental Change, vol. 37, pp. 77–113. Heidelberg: Springer-Verlag. Discusses this issue in more detail.

13. Some people even dislike the phrase "winners and losers" since it suggests some might have vested interests in global change that would make negotiated agreements to slow

down such changes even tougher. Vice President Albert Gore has expressed this view, as I recount from some personal experiences in 1988 on pp. 257–59 of S. H. Schneider. 1990. *Global Warming: Are We Entering the Greenhouse Century?* New York: Vintage.

14. W. D. Nordhaus. Jan.-Feb. 1994. Expert opinion on climatic change. *American Scientist* 82: 45–51.

15. Frosch's calculations are in National Research Council: National Academy of Sciences. 1991. Policy implications of greenhouse warming. For a fuller debate over geoengineering, see the July 1996 issue of *Climatic Change* devoted to the subject.

16. See my "Mediarology" chapter and Epilogue on this issue in *Global Warming: Are We Entering the Greenhouse Century?* San Francisco: Sierra Club Books, 1990, or P. R. Ehrlich and A. H. Ehrlich. 1996. *The Betrayal of Science and Reason.* Washington: Island Press.

17. J. H. Ausubel. Mar.-Apr. 1996. Can technology spare the Earth? *American Scientist* 84: 166–78.

18. R. Herman, S. A. Ardekani, and J. H. Ausubel. 1989. Dematerialization. In *Technology and the Environment.* J. H. Ausubel and H. E. Sladovich, eds. Washington, D.C.: National Academy Press, pp. 50–69.

19. P. R. Ehrlich and A. H. Ehrlich. 1990. *The Population Explosion.* New York: Simon and Schuster.

20. N. Myers. 1993. *Ultimate Security: The Environmental Basis of Political Stability.* New York: Norton. Characterizes (and cites) Myers's long record of looking at trends he sees as potentially threatening, applies theoretical reasoning to project future changes, and suggests solutions. Contrast this with his debate partner, Julian Simon, in their joint book: N. Myers and J. Simon. 1994. *Scarcity or Abundance: A Debate on the Environment.* New York: Norton.

21. M. G. Morgan and D. W. Keith. 1995. Subjective judgments by climate experts. *Environmental Science and Technology* 29: 468–76 A.

22. D. Victor and J. Salt. 1994. From Rio to Berlin: Managing climate change. *Environment* 36: 6–15, 25–32. The authors discuss current negotiations to fashion an international climate convention. The concept of planetary bargaining in this context has its roots in the ideas of the diplomat Harlan Cleveland. In July 1996, Timothy E. Wirth, U.S. Under Secretary of State for Global Affairs (and former U.S. Senator

from Colorado), stunned the delegates to a Geneva meeting (Second Conference of the Parties, Framework Convention on Climate Change). Wirth built on the famous IPCC 1996 "discernible" line by committing the U.S. to the principle that voluntary actions to reduce greenhouse gas emissions are no longer sufficient. Instead, Wirth argued "that future negotiations focus on an agreement that sets a realistic, verifiable, and binding medium-term emissions target." At long last, planetary bargaining may have begun.